MINIMIZING THE USE OF CHEMICALS TO CONTROL SCALING IN SWRO:

IMPROVED PREDICTION OF THE SCALING POTENTIAL OF CALCIUM CARBONATE

MINIMIZING THE USE OF CHEMICALS TO CONTROL SCALING IN SWRO:

IMPROVED PREDICTION OF THE SCALING POTENTIAL OF CALCIUM CARBONATE

DISSERTATION

Submitted in fulfillment of the requirements of
the Board for Doctorates of Delft University of Technology
and of
the Academic Board of the UNESCO-IHE Institute for Water Education
for the Degree of DOCTOR
to be defended in public
on Wednesday, April 27, 2011 at 12:30 hours
in Delft, the Netherlands

by

Tarek Kamal Abdalla Waly
born in Cairo , Egypt

Master of Science in Sanitary Engineering
UNESCO-IHE Institute for Water Education, The Netherlands

This dissertation has been approved by the supervisors:
Prof. dr. G.L. Amy
Prof. dr. G-J. Witkamp

Members of the Awarding Committee;

Chairman	Rector Magnificus, TU Delft, The Netherlands
Vice-chairman	Rector UNESCO-IHE, The Netherlands
Prof. dr. G.L. Amy	UNESCO-IHE / TU Delft, The Netherlands (supervisor)
Prof. dr. G-J. Witkamp	TU Delft, The Netherlands (supervisor)
Prof.dr. M.D. Kennedy	UNESCO-IHE, The Netherlands
Prof. dr. ir. J.C. Schippers	UNESCO-IHE, The Netherlands
Dr. N. Ghaffour	Middle East Desalination Research Center, Oman
Prof. dr. R. Sheikholeslami	The University of Edinburgh, UK
Prof. dr. ir. D. Brdjanovic	UNESCO-IHE / TU Delft, The Netherlands (reserve)

CRC Press/Balkema is an imprint of the Taylor & Francis Group, an informa business

Published by:
CRC Press/Balkema
PO Box 447, 2300 AK Leiden, The Netherlands
e-mail: Pub.NL@taylorandfrancis.com
www.crcpress.com – www.taylorandfrancis.co.uk – www.balkema.nl
ISBN 978-0-415-61578-5 (Taylor & Francis Group)

Acknowledgments

I would like to express my gratitude to my supervision team Prof. Gary Amy, PhD, Prof. Geert-Jan Witkamp, PhD, MSc, Prof. Jan Schippers, PhD, MSc and Prof. Maria Kennedy, PhD for their shepherding roles and critical comments - all of which have helped sharpen this study.

I gratefully acknowledge the financial support of the Middle East Desalination Research Centre (MEDRC), Oman and the support of Dr. Venkat Reddy and Dr. Noreddine Ghaffour.

My deep gratitude to Rinnert Schurer, MSc and Evides NV for allowing us to conduct the pilot testing at their pilot plant in Zeeland, The Netherlands.

I also acknowledge, with thanks, the two Msc. students namely Saleh Saleh and Ruben Munoz Toro for their valuable contributions. To the UNESCO-IHE laboratory staff, namely Fred Kruis, Don van Galen and Peter Heering and Lyzette Robbemont - thanks for your friendliness and support.

My sincere thanks to the department of municipal water and infrastructure staff members for their help, support and friendly advice during the full length of my PhD study, and in particular to Jan Herman Koster, MSc and our hard working secretary Tanny van der Klis.

Thanks to my colleagues Saeed Bagoth, Sergio Salinas Rodriguez, Loreen Villacorte, Assiyeh Tabatabai, Sabine Lattemann, Victor Yangali, Chang Won Ha, Sung Kyu (Andrew) Maeng for their direct and indirect help during my time at UNESCO-IHE.

To my family whom supported me and sacrificed with all what they can to make this step achievable.

Above all, my heartfelt thanks to Allah for his support and accepting my prayers.

This study was funded by the
**MIDDLE EAST DESALINATION
RESEARCH CENTER**
P.O BOX 21
AL Khwair, P.C. 133
Muscat
Sultanate of Oman
Tel: (968) 24 415 500
Fax: (968) 24 415 541
Website: www.medrc.org

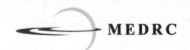

Contents

Summary

Renewable water resources in the Middle East and North Africa (MENA) region are scarce, pushing most of the region's countries far below the water scarcity line of 1,000 m^3/capita/year. This water stress has urged for the implementation of non conventional options to augment water sources. Today, seawater desalination widely used to overcome the water stress in the coastal areas of the MENA region. Although based on the production capacity of fresh water from seawater, distillation process is still the leading technology, seawater reverse osmosis systems (SWRO) is growing rapidly, and expected to dominate the market in the next decade. Research to increase reliability, decrease costs and eliminate the environmental issues associated with concentrate disposal will further strengthen the position of SWRO. This can be achieved partially by optimizing chemicals consumption in the SWRO system. Chemicals in SWRO are used to prevent scaling, particulate/colloidal fouling, biofouling and for system cleaning. Acids and antiscalants are part of these chemicals, and are dedicated to scaling control. As $CaCO_3$ is the most abundant scalant in SWRO systems, acids and/or antiscalnts are frequently used to prevent $CaCO_3$ scaling. The acid and/or antiscalnt doses are calculated using indices e.g, saturation index (SI), saturation ratio (S_a) and Stiff and Davis saturation index (S&DSI). These indices add to the problem by overestimating most likely the scaling potential of the scalant salts which result in overestimating the acid and antisclant doses in SWRO plants. Therefore, accurate prediction of the calcium carbonate scaling potential in the concentrate of SWRO systems is needed to decrease/eliminate the chemicals used for its prevention.

The aim of this research is to improve the accuracy of calcium carbonate scaling prediction in SWRO systems by (i) determining (with the help of calculations and experiments) the degree of saturation of calcium carbonate species and (ii) investigating the kinetics of precipitation of calcium carbonate. As a consequence, it is expected that the need for scaling chemicals such as acid and antiscalant can be reduced / eliminated leading to the production of an environmental friendly concentrate and a decrease in the water production cost.

The kinetics of precipitation were evaluated using induction time experiments in which a designated decrease in pH of 0.03 pH units was considered as an indication of calcium carbonate formation of 0.1 to 0.27 mg/L of $CaCO_3$.
As essential step, the research investigated the potential equilibrium equations and commercial programs used to predict the concentrate pH. The concentrate pH is essential for the determination of the degree of saturation of the concentrate and consequently, the acid and the antiscalant dose.

The effect of salinity on induction time was simulated by using the same calcium and HCO_3^- activities at different ionic strength. The results indicate that there was no effect of ionic strength increase on induction time at the levels of supersaturation (SI = 0.8 to 2.07) evaluated in this study. Furthermore, results showed a negative (linear) correlation between the logarithm of the induction time and the saturation index or the Stiff & Davis saturation index. An investigation for the solubility product value implicitly used in the S&DSI calculations showed that it most likely integrates calcite solubility for saturation calculations for solutions with low ionic strength values (I < 0.06 mole/L) but incorporates vaterite at higher ionic strength levels equal or higher than that of seawater (I > 1.0 mole/L).

In addition, microscopic identification of experimental solution after 24 hours showed that even for very low ionic strength water (I< 0.06 mole/L), mixtures of vaterite and calcite were found in the solution. These results suggest that integrating the solubility product of vaterite, which is 5 times more soluble than calcite, in the saturation index calculations will decrease the estimated supersaturation of seawater concentrate by nearly 0.5 units compared to that using calcite.

Meanwhile, studying of the induction times showed that by testing a wide range of saturations, three different zones were recognized which described homogenous nucleation, heterogeneous nucleation and an intermediate zone with different surface energy value. The nucleation mechanism involved was related to the initial supersaturation of the solution. Homogenous nucleation predominates when the solution is initially supersaturated with regard to the $CaCO_3$ hexahydrated phase. On the other hand, heterogeneous nucleation is the dominant mechanism when the solution is supersaturated with regard to vaterite. In between an intermediate zone exists when the solution is supersaturated with respect to monohydrated $CaCO_3$, but undersaturated with respect to hexahydrated $CaCO_3$.

The essential step in understanding the degree of concentrate supersaturation in SWRO systems is determining the concentrate pH. As a consequence, the concentrate pH was calculated using the CO_2-HCO_3^- and CO_3^{2-}-HCO_3^- equilibrium equations and the results were compared to pH values calculated by two commercial software programs from membrane suppliers and Phreeqc's "evaporation model". Finally, the calculated concentrate pH values were compared to concentrate pH measurements in a SWRO pilot plant. Results showed that concentrate pH was lower than that calculated by equilibrium equations, computer software programs and Phreeqc's "evaporation model" . Firstly, the equilibrium equation employing the CO_2-HCO_3^- equilibrium equation and assuming carbon dioxide is not rejected by the membrane, systematically predicted the concentrate pH value higher than the feed pH. Secondly, the equilibrium equation employing the CO_3^{2-}-HCO_3^- equilibrium equation, systematically predicted the concentrate pH value lower than the feed pH. On the other hand, field measurements showed that the lower the feed pH, the closer the predicted pH using the CO_2-HCO_3^- equilibrium equation to the measured concentrate pH. On the contrary, at high feed pH, the CO_3^{2-}-HCO_3^- equilibrium equation was closer to predict the concentrate pH. Furthermore, at feed pH of 8.0 (normal plant feed pH), the concentrate pH was 7.76.

The effect of inorganic seawater composition, namely Mg^{2+} and SO_4^{2-}, was investigated to determine the effect of these compounds on the kinetics of precipitation of $CaCO_3$. The results of this investigation suggest that the presence of Mg^{2+} and SO_4^{2-} were very effective in hindering or retarding the crystal growth of $CaCO_3$ crystals. Furthermore, environmental scanning electron microscope (ESEM) and X-ray diffraction (XRD) analysis of the crystals formed showed that the presence of Mg^{2+} prevented the transformation of aragonite into the more stable calcite, even after 5 months of crystal ripening. On the other hand, after the same ripening time the final phase formed in the presence of SO_4^{2-} was calcite.

This suggests that studying $CaCO_3$ scaling in SWRO plants without taking into account the role of Mg^{2+} and SO_4^{2-} ions may result in inaccurate scaling prediction, based on induction time measurements and hence an overestimation the probability of scaling and consequently, the acid and/or antiscalant doses.

Finally, the outcome of this study indicates that the pH of concentrates of seawater reverse osmosis plants are lower than commonly expected. This effect is attributed to the effect of ionic strength on the activities of the ions involved and   the mechanisms governing the pH in the feed – concentrate. As a consequence the degree of supersaturation is lower as well.

If vaterite is governing the solubility of calcium carbonate – as indicated in induction time measurements in artificial seawater solutions, containing magnesium and sulphate as well - the saturation of seawater concentrate is expected to be nearly 0.7 SI units.

This opens, together with the rather long induction time measured in artificial seawater with magnesium and sulphate (see chapter 6) the opportunity to reduce or even stop dosing acid/antiscalant used to prevent $CaCO_3$ scaling in SWRO plants. To prove this hypothesis, a SWRO pilot plant in the Netherlands has operated on open seawater at 40% recovery for more than 6 months without any acid/antiscalant and showing no scaling, which indicates that the concentrate is either undersaturated with respect to $CaCO_3$ or has very slow precipitation kinetics.

Samenvatting

Herbruikbare waterbronnen zijn schaars in de regio Midden-Oosten en Noord-Afrika, waardoor het grootste gedeelte van de regio ver beneden de waterschaarstelimiet van 1000 m^3 per persoon per jaar leeft. Deze waterschaarste heeft ertoe geleid dat er op zoek gegaan moest worden naar non-conventionele opties om de bestaande watervoorraden aan te vullen.

Vandaag de dag is ontzouting van water een veelgebruikte techniek om de waterschaarste het hoofd te bieden in de kustgebieden van het Midden-Oosten en Noord-Afrika. Hoewel het destillatieproces nog steeds de voornaamste techniek is om drinkwater uit zeewater te produceren wint de omgekeerde osmose van zeewater (SWRO) aan populariteit. In de komende 10 jaar wordt verwacht dat SWRO de marktleidende technologie gaat worden.

Onderzoek gericht op het vergroten van de betrouwbaarheid, het verlagen van de kosten en het elimineren van negatieve milieu-effecten zullen bijdragen aan een sterke marktpositie van SWRO. Dit kan deels bereikt worden door het chemicaliënverbruik te optimaliseren.

Chemicaliën worden ingezet tijdens het ontzoutingsproces om scaling, fouling en biofouling te voorkomen, en om het systeem te reinigen. Zuren en anti-scalants zijn voorbeelden van deze chemicaliën die ertoe dienen scaling in de hand te houden. Omdat calciumcarbonaat een van de voornaamste veroorzakers is van scaling, wordt er vaak gericht gedoseerd op het voorkomen van CaCO3 vorming. De te doseren hoeveelheden worden berekend aan de hand van indices, bijvoorbeeld de verzadigingsindex (SI), de verzadigingsratio (S_a) en de Stiff en Davis verzadigingsindex (S&DSI). Doordat deze indices de oververzadiging vaak overschatten, wordt er dikwijls te veel anti-scalant gedoseerd. Een precieze voorspelling van de mate van oververzadiging is noodzakelijk om de dosering van anti-scalants te reduceren of zelfs overbodig te maken.

Het doel van dit onderzoek is om de precisie te verhogen waarmee calciumcarbonaatvorming voorspeld kan worden in ontzoutingssystemen voor zeewater middels: (i) het bepalen van de verzadigingsgraad van verschillende calciumcarbonaatvormen (met behulp van berekeningen en experimenten) en (ii) het onderzoeken van de precipitatiekinetiek van calciumcarbonaat. Verwacht wordt dat de benodigde hoeveelheid anti-scalants en zuren sterk zal afnemen waardoor het concentraat, wat overblijft na het ontzouten van zeewater, op een milieuvriendelijke manier verwerkt kan worden en de kostprijs van ontzout zeewater daalt.

De precipitatiekinetiek is onderzocht middels inductietijd-experimenten waarbij een afname van 0.03 pH eenheden beschouwd werd als een indicatie van de productie van 0.1 tot 0.27 mg/L $CaCO_3$. Een essentiële stap in dit proces was het testen van verschillende commerciële computerprogramma's om de pH te kunnen voorspellen. De pH-waarde van het concentraat is van belang voor de bepaling van de verzadigingsgraad en daarmee voor de benodigde hoeveelheid anti-scalant. Het effect van de saliniteit op de inductietijd werd gesimuleerd door gebruikmaking van gelijkblijvende activiteiten van calcium en HCO_3^- bij verschillende ionensterktes. Uit de resultaten bleek dat de ionensterkte geen effect heeft op de inductietijd in het bereik van de onderzochte oververzadigingswaarden (SI = 0.8 to 2.07). Verdere analyses lieten een negatieve (lineaire) correlatie zien tussen de logaritme van de inductietijd en de saturatie-index of de Stiff en Davis verzadigingsindex.

Nader onderzoek naar het oplosbaarheidsproduct, welke indirect wordt gebruikt in de S&DSI berekeningen, liet zien dat zeer waarschijnlijk de oplosbaarheidswaarde voor calciet genomen wordt voor lagere ionensterktes (I < 0.06 mol/L), terwijl bij ionensterktes gelijk aan of hoger dan die van zeewater (I > 1.0 mol/L) de waarde voor valeriet gebruikt wordt. Microscopisch onderzoek toonde aan dat zowel calciet als valeriet aanwezig is in de oplossing, zelfs bij zeer lage ionensterktes (I< 0.06 mol/L). Wanneer het oplosbaarheidsproduct van valeriet, dat vijf maal hoger is dan dat van calciet, gebruikt wordt in verzadigingsberekeningen, dan wordt de mate van verzadiging met bijna 0.5 eenheden onderschat.

Inductietijdsexperimenten uitgevoerd bij een breed scala aan verzadigingswaarden duidden op drie zones die beschreven kunnen worden als homogene nucleatie, heterogene nucleatie, en een intermediaire zone, met een afwijkende waarde voor de oppervlakte-energie.
Het kernvormingsmechanisme is gerelateerd aan de mate van oververzadigings ten tijde van het begin van het experiment. Homogene nucleatie vindt plaats indien de oplossing initieel oververzadigd was voor hexagehydrateerd $CaCO_3$. Is de oplossing verzadigd met betrekking tot vateriet, dan vindt heterogene nucleatie plaats. Daartussenin bestaat een zone waarin de oplossing oververzadigd is met betrekking tot vateriet maar onderverzadigd voor hexagehydrateerd $CaCO_3$.

Het effect van de samenstelling van zeewater, met name de aanwezigheid van Mg^{2+} en SO_4^{2-}, is onderzocht om uitsluitsel te krijgen over het effect van deze componenten op de kinetiek van calciumcarbonaatprecipitatie. De resultaten laten zien dat de aanwezigheid van Mg^{2+} and SO_4^{2-} zeer effectief is ter verhindering of vertraging van de groei van calciumcarbonaatkristallen. Environmental scanning electron microscope (ESEM) en X-ray diffraction (XRD) analyses toonden aan dat in aanwezigheid van Mg^{2+} de gevormde aragonietkristallen niet overgingen in de stabielere vorm calciet, zelfs niet na 5 maanden rijping. In de aanwezigheid van SO_4^{2-} bleken de kristallen wel over te gaan naar calciet binnen deze tijd. Hieruit volgt dat het verwaarlozen van het effect van Mg^{2+} en SO_4^{2-} ionen bij het bestuderen van calciumcarbonaat-scaling in SWRO installaties leidt tot onjuiste voorspellingen: de waarschijnlijkheid van scaling wordt overschat en daarmee de ook de benodigde hoeveelheid zuur en / of antiscalant.

Tot slot kunnen we stellen dat de resultaten van dit onderzoek laten zien dat de pH van de concentraten van omgekeerde osmose-installaties voor zeewater lager is dan normaal gesproken wordt verwacht. Dit is toe te schrijven aan het effect van de ionensterkte op de betrokken ionen en de mechanismen die de pH waarde bepalen in de instroom en het concentraat. Als gevolg daarvan is de mate van oververzadiging ook lager. Indien vateriet bepalend is voor de oplosbaarheid van calciumcarbonaat, zoals wordt gesuggereerd door de metingen aan de inductietijdsexperimenten in kunstmatige zeewateroplossingen die ook magnesium- en sulfaationen bevatten, dan bedraagt de oververzadiging van het zeewaterconcentraat bijna 0.7 verzadigingseenheden.

Dit, samen met de relatief lange inductietijd gemeten in de experimenten met kunstmatige zeewateroplossingen met magnesium en sulfaat (zie Hoofdstuk 6), biedt mogelijkheden om de dosering van zuur en anti-scalant ter voorkoming van $CaCO_3$ scaling in SWRO installaties te reduceren of zelfs te stoppen. Om deze hypothese te bewijzen heeft een SWRO proefinstallatie in Nederland gedurende 6 maanden gedraaid met open zeewater en 40% terugwinning zonder toevoeging van zuur of anti-scalant, zonder enige scaling op de membranen. Dit toont aan dat het concentraat onderverzadigd is met betrekking tot $CaCO_3$, of dat de precipitatiekinetiek bijzonder langzaam is.

Chapter 1

Introduction

Background

Based on United Nations population projections, the number of people in water-scarce countries increased from 130 million in 1990 to between 650 and 900 million in 2010, and is expected to rise to 1800 million people by 2025 [1-3]. According to world bank definition, a country is considered to suffer from water scarcity when the renewable water resources in such a country are less than 1000 m^3/capita/year [2, 4]. Today more than 20 countries suffer from water scarcity, most of which are located in the Middle East and North Africa region (MENA). Furthermore, water shortage in the MENA region is expected to become more acute due to an increasing demand for water and multiple-source pollution [5].

Being partly arid and partly semi-arid, the MENA region has one of the world's lowest rainfalls, mostly seasonal and with erratic distribution [6]. Renewable water resources in the MENA region are scarce, pushing most of the region's countries far below the water scarcity line of 1,000 m^3/capita/year (Figure 1) [4, 7]. Moreover, the non-renewable water sources, mostly in groundwater basins, are being exhausted at an increasing rate by extensive irrigation [8].

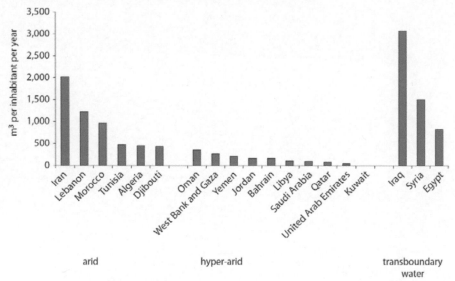

Figure 1: Renewable water per capita for the MENA countries (m^3/capita/year) [7]

Limited available water resources are forecast to pose a great challenge to achieving the ambitious levels of development in the MENA region [4, 8-9]. Moreover, in the Gulf countries where 80% of the population are foreign workers, water scarcity may cause reverse immigration and lead to an economic downturn [9]. In the MENA region, the water stress has pushed for the implementation of non conventional options to augment water sources as water saving, water harvesting, waste water reuse, water transportation and desalination [4].

Unlike water harvesting, water reuse and transportation are more feasible options when it comes to sustain the need of large communities. In the MENA region water reuse is a slowly growing industry due to religious/cultural constrains and public trust in the produced water quality. On the contrary, seawater desalination is a rapidly growing coastal-based industry worldwide and in the MENA region [10]. The combined production capacity of all seawater desalination plants worldwide has increased by 30% over the last two years: from 28 million cubic meters per day in 2007—which is the equivalent of the average discharge of the River Seine at Paris—to more than 36 million cubic meters per day in 2009 [6]. Water desalination techniques can be broadly divided into four categories based on the driving force: *e.g.,* thermal (based on temperature differences), pressure, electrical potential, or chemical potential-driven processes (Table 1) [11-12]. The thermally-driven processes are the oldest and the most widely used for seawater desalination, especially that using multi-stage flash evaporation (MSF) and multi-effect desalination (MED) methods [6, 13].

The pressure-driven process, involving reverse osmosis (RO) membranes, is taking an increasing share of the world desalination capacity [14]. Since 2004, and based on the production capacity, reverse osmosis of brackish water and seawater has been the leading technology compared to the distillation processes. This situation is reversed for seawater alone where distillation processes are still the market leader with a total installed capacity of 62% [13].

Table 1: Desalination methods and the corresponding process description[13]

Separation	Driving Force	Process	Desalination method
Water from Salt	Thermal	Evaporation	Multi-Stage Flash
			Multi Effect Distillation
			Thermal Vapour Compression
			Solar Distillation
		Crystallisation	Freezing
			Gas Hydrate Processes
		Filtration/Evaporation	Membrane Distillation
	Pressure	Evaporation	Mechanical vapour Compression
		Filtration	Reverse Osmosis
Salt from Water	Electrical	Filtration	Electrodialysis
	Chemical	Exchange	Ion Exchange

The use of chemicals in SWRO (Table 2), *e.g.* to control biofouling (biocides, chlorine, chlorine dioxide, chloramines and DBNPA), particulate fouling (coagulant and coagulant aid) and scaling (acids and antiscalants) is a permanent concern as part of the chemicals are concentrated in the brine, resulting in an environmental impact during disposal [6, 15-21]. Some of the concentrated disposed chemicals have been suggested to have an environmental consequence when discharged to the environment. For instance, the water body in the Gulf area alone receives nearly 60 tons of antiscalants every day which may interact with the natural dissolved metal ions in the seawater body and disturb the balance of aquatic life [22].

Minimization or even elimination of one or more chemical dosages in SWRO has interesting advantages *e.g.* decreasing the amount of chemicals discharged to the environment and hence reducing the concentrate environmental impact for the concentrates disposal [23], saving capital and chemical costs [24], simplifying equipment and storage facilities, and simplifying operation and maintenance [25].

Table 2: Types of chemicals frequently used in SWRO system

	Chemical	Dose (mg/L)	Use
Pre-treatment	Chlorine, biocide	1-2	Control growth
	Sodium bisulphite	3-6	Neutralize chlorine
	Coagulant	1-1.5	Flocculation
	Coagulant aid	0.5-1	Flocculation
	Acid	20-55	Prevent alkaline scaling
	Antiscalants	1-4	Preventing scaling
Cleaning chemicals	Acid	Shock dose	Remove scaling
	Alkali	Shock dose	Remove particulate fouling
	Biocides	Shock dose	Remove bacteria
Post-treatment	Lime/limestone	20-70	Control corrosiveness and water conditioning
	Post Chlorination	0.3-0.5	Disinfection

Minimizing the SWRO chemical consumption can be achieved partially by decreasing the chemicals used for scaling control in SWRO system. The problem of scaling arises when the feed water is converted into product, and as a consequence dissolved compounds exceeds their saturation limits in the concentrate. This results in salt crystallization/precipitation on the membrane surface and loss of productivity, increase of driving pressure or even membrane damage.

In SWRO systems, calcium carbonate is considered the most significant potential scalant. The scaling tendency of $CaCO_3$ is assessed using different indices, e.g., supersaturation ratio (S_a), supersaturation index (SI) and the semi-empirical Stiff and Davis saturation index (S&DSI) [26-28]. S_a and SI calculate the saturation by correlating the activities of the reacting ions to the solubility product of the crystalline phase. In the case of $CaCO_3$, the calcite phase solubility is used. On the other hand, the widely used S&DSI (introduced in 1952) was based on laboratory experiments to determine the scaling limit of $CaCO_3$ in NaCl solutions at different ionic strength and temperature ranges [27]. One of the main disadvantages of these indices is, that they can not predict the period of metastability that supersaturated solutions exert before the start of the nucleation process. This period is referred to in literature as 'induction time' [29]. Based on the degree of supersaturation, this period can be long enough that the supersaturated concentrate will start to precipitate after its disposal.

Acids and/or antiscalants are frequently used to prevent calcium carbonate scaling in SWRO systems [30]. The chemical consumption is based on the degree of the concentrate saturation with respect to $CaCO_3$ and doses in the feed are designed to keep the concentrate undersaturated [30]. The high consumption of chemicals for scaling control in SWRO may be attributed to overestimation of the degree of saturation of $CaCO_3$ in the concentrate. This overestimation in the degree of saturation may be a result of using calcite as the critical phase in the saturation calculations using the saturation ratio or the saturation index. The use of other phases, e.g., vaterite which is 8 times more soluble than calcite will result in a much lower saturation than previously anticipated. Moreover, S&DSI was developed in NaCl solutions and not in real seawater medium [27] where inorganic and organic ions may affect the activity of the reacting ions and hence decrease the overall solution saturation.

Increasing the accuracy of these indices depends on (i) which phase of $CaCO_3$ is precipitating first, (ii) correct concentrate pH prediction and (iii) taking into account the effect of the ionic interaction between other ions present in seawater, on the availability of $CaCO_3$ to precipitate. These various factors may result in $CaCO_3$ scaling control reduction/elimination, leading to an overall improvement in the sustainability of SWRO plants.

1.1 Problem Definition

Minimization/elimination of the chemicals used for $CaCO_3$ scaling prevention may have an important role in future system design, where environmental regulations become stricter and water production costs need to be decreased. Moreover, elimination/reduction of acids and/or antiscalants results in an environmental friendly concentrate which is much safer to dispose of. In addition the chemical-free concentrate - depending on land availability - could be directed to solar lagoons for salt production and hence increase the economic benefit of the plant. As SWRO plants use either acids or antiscalants in $CaCO_3$ scaling control, an illustrative example for a medium size plant (20,00 m3/d) shows the reduction in the annual running cost for removing the acids or the antiscalants (Table 3)

Table 3: Acid and antiscalants cost for pre-treatment of 20,000 m^3/d SWRO plant working at 40% recovery

	Dose ppm [6]	Feed flow	Raw chemicals Kg/d	Average Dose kg/d	Cost per US$/kg [31]	Yearly cost US$	Cost US$/$m^3$
pH correction (sulphuric acid)	20-50	50,000	1,000-2,400	1,600	0.1-0.55	58,000-321,000	0.008-0.04
Antiscalant	2-4	50,000	100-200	150	1.0-3.6	55,000-197,000	0.0075-0.03

*The cost reduction varies to a great extent with the price and quality of antiscalants and acids used.

Furthermore, the unit costs for any desalination plant are considered an important factor in the feasibility studies of SWRO plants. Chemicals comprise around 5-10% of the operational cost of any SWRO plant with the majority of this amount located in the chemical pre-treatment [24, 30, 32]. The chemical dosing systems, *e.g.*, dosing pumps, tanks, mixers, and storage facilities comprise much less in term of capital cost, about 4-6% [33-34]. In the case of acids or antiscalants elimination in medium capacity plants (Table 3), the plant running cost can be annually decreased by 0.015-0.07 US$/m^3 of the total production cost. This large range of cost reduction is attributed to the great difference in market value and quality of acids and antisclants used in SWRO systems.

1.2   The overall aim of the research

The aim of this research is to improve the accuracy of calcium carbonate scaling prediction in SWRO systems by (i) determining (with the help of calculations and experiments) the degree of saturation of calcium carbonate species and (ii) investigating the kinetics of precipitation of calcium carbonate. As a consequence, it is expected that the need for scaling chemicals such as acid and antiscalant can be reduced / eliminated leading to the production of an environmental friendly concentrate and a decrease in the water production cost.

1.3   Research objectives

Calcium carbonate scaling is considered the most significant potential scaling salt in SWRO systems. Studying the accuracy of calcium carbonate scaling prediction in SWRO systems may lead to further SWRO system optimization. The project objectives are divided among the thesis chapters (2-7). Each chapter investigates the effect of different parameters affecting the kinetics of $CaCO_3$ formation.

Chapter 2 The main objective of this chapter is to develop a robust, accurate, online and sensitive method to measure the induction time of calcium carbonate in a high saline medium (similar to SWRO concentrates).

Chapter 3 The objective of this chapter is to investigate the effect of particles on the induction time of calcium carbonate. Particles are naturally present in seawater with different sizes and composition and hence are expected to influence the induction time by providing more surface area for nucleation. For the fulfilment of the chapter objective, the effect of foreign particles surface area on the induction time was explored.

Chapter 4 The objective of this chapter is to investigate the effect of salinity on the kinetics of $CaCO_3$ formation in synthetic seawater concentrate at 30% and 50% recovery rates. Furthermore, the study aims to determine the first precipitating phase of $CaCO_3$. The results from chapter 3 and 4 will be used to develop a prediction model for induction time in saline medium.

Chapter 5 The main objective of this chapter is to predict the concentrate saturation by predicting the concentrate pH. The pH calculation is based on $CaCO_3$ system equilibrium equations to predict the concentrate pH when the feed pH is known. The developed calculations will be verified by field measurements from a real SWRO pilot.

Chapter 6 The main objective of this chapter is to investigate the effect of inorganic ion presence, namely magnesium and sulphate, on the driving force for nucleation and crystal growth of $CaCO_3$ through measuring the induction time in synthetic seawater concentrate at 30% and 50% recovery rates.

Chapter 7 contains the general conclusions for this thesis and recommendations for future work.

1.4   References

1.   Morrison, J., et al., Water scaricity and climate changes: Growing Risks for businesses & investors, in Ceres. 2009, Pacific institute: Boston.
2.   FAO, Coping with water scaricity challange of the twenty first century, in world water day U. water, Editor. 2007.
3.   Forstmeier, M., et al., Feasibility study of wind powered desalination. Desalination, 2007. 203: p. 463-470.
4.   Hamdy, A., Water Crisis and Food Security in the Arab World: The Future Challenges, A.W. Council, Editor. 2008: Beirut.
5.   Report, H.D., Water Governance Challenges: Managing Competition and Scarcity for Hunger and Poverty Reduction and Environmental Sustainability, H.D.R. Office, Editor. 2006, Stockholm International Water Institute: Stockholm.
6.   Lattemann, S., Development of an environmental impact assessment and decision support system for seawater desalination plants. 2010, UNESCO-IHE: Delft.
7.   World-bank, Making most of the scaricity: acountability for better water management results in the Middle East and North Africa, in MENA development report. 2007, World bank: Washinton DC.
8.   Amer, K.M. and A. Abdel-Wahab, Water and Human Development in Qatar: Challenges and Opportunities, in Qatar National Vision 2030. 2009: Doha.
9.   Malki, A.S.A., Business Opportunities in Water Industry in Qatar, W.n. affairs, Editor. 2008, Qatar General Electricity and Water Corp. (KAHRAMAA): Doha.
10.   Hochstrat, R., et al., Options for water scarcity and drought management - the role of desalination, in Desalination for the environment-clean water and energy, EDS, Editor. 2009: Baden-Baden.

11.   Awerbuch, L., ed. The status of desalination in today's world. Desalination and Water Re-Use, ed. S. Nicklin. 2004, Tudor Rose: United Kingdom.

12.   M. Ahmad, P.W., Assessment of desalination technologies for high saline brine applications - discussion paper, in Desalination for the environment-clean water and energy, EDS, Editor. 2009: Baden-Baden.

13.   Trieb, F., Concentrating solar power for seawater desalination, in AQUA-CSP, G.A.C. (DLR), Editor. 2007, Institute of Technical Thermodynamics: Stuttgart.

14.   Lattemann, S. and T. Bleninger, Seawater Desalination and the Environment, B. Group, Editor. 2010: Jeddah.

15.   Alhadidi, A., Process evaluation of the Klazienaveen Reverse Osmosis (RO) Plant for the Water Supply Company, Drenthe (WMD), in MWI 2006, UNESCO-IHE: Delft.

16.   Waly, T., et al., Will calcium carbonate really scale in seawater reverse osmosis? Desalination and water treatment, 2009. 5: p. 252-256

17.   Waly, T., et al., Reducing the calcite scaling risk in SWRO: role of Mg^{2+} & SO_4^{2-}, in IDA world congress: Desalination for a better world. 2009: Dubai.

18.   Al-Rawajfeh, A., H. Glade, and J. Ulrich, Scaling in multiple-effect distillers: the role of CO_2 release. Desalination, 2005. 182: p. 209-219.

19.   Amy, G., Fundamental understanding of organic matter fouling of membranes. Desalination, 2008. 231: p. 44-51.

20.   Chen, T., A. Neville, and M. Yuan, Assessing the effect of Mg^{2+} on $CaCO_3$ scale formation—bulk precipitation and surface deposition. Journal of Crystal Growth, 2005. 275: p. 1341-1347.

21.   De Yoreo, J.J., A. Wierzbicki, and P.M. Dove, New insights into mechanisms of biomolecular control on growth of inorganic crystals. Crystal Engineering Comm., 2007. 103: p. 19237-19242.

22.   UNEP, Desalination resource and guidance manual for environmental impact assessments, S. Lattemann, Editor. 2008, United Nations Environment Programme, Regional Office for West Asia, Manama, and World Health Organization, Regional Office for the Eastern Mediterranean: Cairo.

23.   Lattemann, S. and T. Höpner, Impacts of seawater desalination plants on the marine environment of the Gulf Earth and Environmental Science 2008. DOI: 10.1007/978-3-7643-7947-6_10 p. 191-205.

24.   Reddy, K.V. and N. Ghaffour, Overview of the cost of desalinated water and costing methodologies. Desalination, 2007. 205: p. 340-353.

25.   Ning, R. and J. Netwig, Complete elimination of acid injection in reverse osmosis plants. Desalination, 2002. 143: p. 29-34.

26.   Sheikholeslami, R., Assessment of the scaling potential for sparingly soluble salts in RO and NF units. Desalination, 2004. 167: p. 247-256.

27.   Stiff, H.A. and L.E. Davis, A method for predicting the tendency of oil field waters to deposit calcium carbonate. Petroleum transactions, 1952. 195: p. 213-216.

28.   Wilf, M., The guidebook to membrane desalination technology. 2007, L'Aquila: Desalination publications.

29.   Sohnel, O. and J. Garside, Precipitation basis principals and industrial applications. 1992, Oxford: Butterworth-Heinemann.
30.   Li, N., et al., Advanced membrane technology and applications. 2008, New Jersey: John Willey & Sons.
31.   Malki, M., Case study: optimizing scale inhibition costs in reverse osmosis desalination plants. 2009, American Water Chemicals, Inc. p. 1-8.
32.   Borsani, R. and S. Rebagliati, Fundamentals and costing of MSF desalination plants and comparison with other technologies. Desalination, 2005. 182(2005): p. 29-37.
33.   Hafez, A. and S. El-Manharawy, Economics of seawater RO desalination in the Red Sea region, Egypt. Part 1. A case study. Desalination, 2003. 153(1-3): p. 335-347.
34.   Gill, J., A Novel inhibitor for scale control in water desalination. Desalination, 1999. 124: p. 43-50.

Chapter 2

Development of a reliable CaCO₃ induction time measuring tool: the role of salinity and CO₂

2.1. Abstract

Induction time measurements can be used to determine the kinetics involved in the precipitation of $CaCO_3$. The longer the induction time, the higher the probability that $CaCO_3$ will not precipitate inside the SWRO system. Induction time can be measured in several ways, e.g., monitoring calcium by measurements with inductively coupled plasma (ICP), conductivity or pH. Although various methods have been used to monitor induction time, these methods are not always directly applicable in high salinity concentrates. In this research study, the accuracy and sensitivity of measuring the induction time of $CaCO_3$ in low (0.063 mole/L) and high (1.34 mole/L) ionic strength concentrates using ICP measurements, conductivity and pH were investigated.

Two different methods were used in investigating the induction time by monitoring the change in Ca^{2+} concentration by ICP. The first used the filtrate and measurement of the concentration of the Ca^{2+} ions. The second was to measure the crystals retained after filtering the sample through a 0.2 µm filter, after which, the filters were acidified and calcium measured in the solution with ICP.

For pH measurements a new highly accurate and stable pH meter suitable for high salinity solutions was used.

In addition, the difference between an open and closed reactor on of the induction times was investigated

The measurement of calcium in solution with ICP, to monitor induction times, turned out to be rather inaccurate. Determining the amount of precipitated calcium carbonate by filtering the solution and measuring the retained crystals by dissolving and ICP suffered from high blank values. These high blanks are attributed to remaining calcium, present in the solutions and in the filters

Conductivity measurements were inaccurate as well, since the change in conductivity due to precipitation of calcium carbonate was too low to be detected accurately.

Measurements with a new (on the market) highly sensitive and stable pH meter were very accurate. As a consequence this method has been selected to be applied for measuring induction times in this study.

Open reactors showed obviously a significant exchange of carbon dioxide with the atmosphere resulting large deviations in observed induction times in closed reactors. Consequently closed reactors were applied in all experiments for this study.

Keywords: Induction time measurements, method, calcium carbonate, seawater reverse osmosis

2.2. Background

Introduction

Scaling is considered a major constraint in the design and operation of SWRO. Calcium carbonate, as the most abundant scalant in SWRO, is commonly assumed to precipitate immediately when its solubility is exceeded [1-4]. In practice, supersaturated solutions exert a degree of stability before the start of precipitation (induction time) [1-2, 4, 6-13]. The longer the induction time the higher the possibility that precipitation will not occur on the membrane surface. This part of research aims to develop a reliable yet simple method to measure the induction time in high ionic strength waters similar to that of seawater concentrates.

Induction time

Theoretically, once supersaturation is achieved, precipitation is possible. However supersaturation is not the only factor that is involved in scale formation, which is also influenced by precipitation kinetics [3]. The kinetics in general describes the period of stability a supersaturated solution goes through before crystallization starts (Figure 1) [3, 5]. In this period ions start clustering as proto-nuclei of up to 1000 atoms in a readily reversible reaction with proto-nuclei forming and disbanding. As the proto-nuclei grow, the ions begin aligning in an 'orderly' way with the formation of stable nuclei. This stage of crystal development is also reversible, but as the nuclei grow, reversal becomes less likely. The final stage is the irreversible growth of crystals from nuclei. Once formed, the crystals can grow indefinitely as long as the respective salt exceeds its solubility product [6]. In practice the period of metastability preceding the precipitation process is commonly indicated as the induction time (t_{ind}) [3, 5, 7-8].

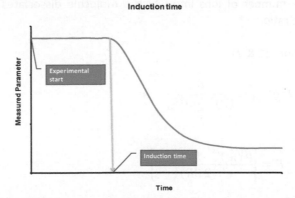

Figure 1: A schematic diagram showing the induction time measurements where the Y-axis shows the measured parameter, e.g,. pH and the X-axis shows the time; the graph shows the period of stability of a supersaturated solution before nucleation takes place

Although there is no direct relation between scaling in SWRO systems and induction time, an induction time longer than the average retention time of concentrate in a single stage SWRO may be an indication that scaling will probably not occur on the membrane surface [5]. This hypothesis may be valid for a clean virgin membrane but is less true when there is profound fouling on the membrane surface. Fouling may hinder the flow on the feed spacers and considerably increase the retention time to more than the estimated values [9-11].

Experimentally, it is very difficult to determine the formation of the first nuclei, and consequently, a part of the induction time measured in experiments may also include growth to a detectable size [3]. The induction time calculation depends on the relation between the growth time (t_g) and the nucleation time (t_n). If the assumption that the growth time to a detectable size is quite short (depending on the measuring technique) and the largest portion of induction time is that for nucleation $(t_n >> t_g)$, then the steady state nucleation rate can be related to the induction time as shown in equations 1-7 [3].

$$t_{ind} \propto J^{-1}$$
1.

Where the nucleation steady state rate can be expressed in

And the rate of nucleation in such cases can be written as follows:

$$J = \Omega \ exp\left(\frac{-\beta\vartheta^2\gamma_s^3}{k_bT\varnothing^2}\right)$$
2.

Where β is the geometric factor, ϑ Molecular volume, γ_s is the surface energy, Ω Pre-exponential factor, T is the temperature in Kelvin and k_b is the Boltzmann constant

$$\varnothing = vk_bTLnS_a$$
3.

Where υ is the number of ions into which a molecule dissociates and S_a is the supersaturation ratio.

Based on equations (1 & 2)

$$logt_{ind} = \frac{B}{(T)^3.log^2S_a} - A$$
4.

Where

$$B = \left[\frac{\beta V_m^2.\gamma^{s^3}.f(\theta).N_A}{v^2.(2.3R)^3}\right]$$
5.

In equation 5, The value of ƒ(θ) determines the nucleation mechanism. Homogeneous nucleation is recognized by ƒ(θ) =1 while for heterogeneous nucleation ƒ(θ) <1. Values as low as 0.01 were reported for ƒ(θ) [3]

And

$$A = log\Omega \hspace{4cm} 6.$$

Published experimental data are available to show that the relation between log (t_{ind}) and $T^{-3}Log^{-2} S_a$ for a large number of substances e.g. $CaCO_3$, $BaSO_4$, $SrSO_4$, and $BaCrO_4$, is linear if the range of supersaturation investigated is limited [5, 7, 12-17].

If short range of supersaturation is covered the induction time relation is simpler and a linear fit can be found between the logarithm of the induction time and that of the saturation ratio [3, 18].

$$logt_{ind} = k - n.log(S_a) \hspace{3cm} 7.$$

In such cases n represents the reaction order while k has no physical meaning and can be determined experimentally [3, 19].

It has been observed that temperature variation and mixing speed affected the precipitation kinetics, e.g. the faster the mixing the shorter the induction time. [20-21]. Moreover, the mixing technique, whether mechanical, magnetic stirrers or gas bubbling, was noted to affect the induction time , the maximum solubility and the crystal size formed [20, 22].

Measurement tools

There are various methods used to measure the induction period as shown in Table 1. The most common methods used are calcium ion specific electrode, conductivity meter, pH meter, inductively coupled plasma (ICP), (laser) turbidity meter and UV absorbance. Generally, the faster the method's ability to detect the first formed crystals, the more accurate the method is in determining the induction time.

Table 1: Common methods used in induction time measurements

Method	Reference
Turbidity meter/light scattering	[13, 23-24]
Decrease membrane permeability	[25-27]
Calcium electrode	[16, 28-30]
ICP	[5, 31-36]
UV spectrophotometer	[34, 37]
Conductivity meter	[7, 14, 38-39]
Particle counter/ microscope detection	[15, 17, 22, 40]
Quartz microbalance technique	[41-44]
pH	[37-38, 45-50]
Visual	[7]

The drop in the calcium concentration due to calcium carbonate precipitation was frequently used for measuring the induction time of calcium salts. The calcium measurements were carried out either by calcium specific electrode or ICP. The ICP was preferred due to its high accuracy and its very low detection limit (10µg/L) which enabled the detection of the first formed crystals. This technique was claimed by other researchers to be ideal for induction time measurements for SiO_2, $CaCO_3$ and $CaSO_4$ for salinities up to SWRO concentrates [5, 31-36].

Conductivity measurements are widely used in induction time experiments [7, 14, 38-39]. The measurement depends on recognizing the decrease in water conductance due to precipitation of the desired salt. The main advantage of this technique lies in its simplicity and the absence of further analytical measurements to obtain the induction time values. The same concept applies to the pH measurements, with the limitation that these measurements are only to be used when the precipitating salts cause a change in the hydrogen ion concentrations [37-38, 45-50]. The pH measurements can be taken directly by using a pH electrode [21].

Effect of exchange CO_2 with atmosphere on induction time

The $CaCO_3$ induction time experimental procedures in the literature have normally marginalized the effect of the gas-liquid interface in their reactor design [7-8, 20, 32, 38, 47-48, 51-53]. One raised the claim that the reaction is very fast for a significant exchange of CO_2 with the solution [7-8, 36, 54]. Induction time measurements have been, in most cases, carried out in open reactors while mixing.

Although the distribution of carbon species in closed and open systems is extremely different (Figures 2 & 3), a few comparisons between the effect of eliminating all gas contact with water and its presence on the induction time were elucidated. One of these comparisons was made by Morales et al, 1996 [49] where there were differences between the induction time experimental work done in the absence and the presence of atmospheric CO_2, where the latter directly affected the experimental pH measurements (equations 8-13).

$$Ca^{2+} + 2HCO_3^- \leftrightharpoons CaCO_3 + CO_2 \uparrow + H_2O \qquad\qquad 8.$$

And

$$CO_2(aq) + H_2O \leftrightarrow H_2CO_3 \qquad\qquad 9.$$

Where at equilibrium the amount of H_2CO_3 is always constant and as large as 10^{-3} of the CO_2 in water [55-56].

$$CO_2(aq) + H_2O \leftrightarrow HCO_3^- + H^+ \qquad\qquad 10.$$

$$HCO_3^- + H^+ \leftrightarrow CO_3^{2-} + H_2O \qquad\qquad 11.$$

Traditionally relations are used in which the dissociation constants depend on salinity.

$$pH = pK_{a1} + log\frac{[HCO_3^-]}{[CO_2]} \qquad\qquad 12.$$

$$pH = pK_{a2} + log\frac{[CO_3^{2-}]}{[HCO_3^-]}$$ 13. .

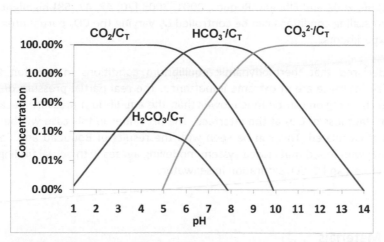

Figure 2: Relation between different inorganic carbon species with respect to solution pH in a closed system with the assumption of a total inorganic carbon content of 10^{-3} mole/L.

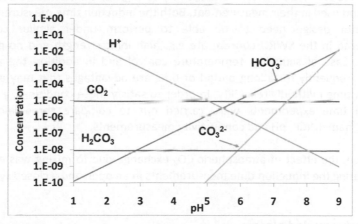

Figure 3: Relation between the concentrations of different inorganic carbon species at constant $[CO_2]$ content at various pH values in an open system with the assumption of a total inorganic carbon content of 10^{-3} mole/L.

Simulating closed reactors conditions was recently carried out by pressurizing N_2 [57] or CO_2 inside a vessel to prevent atmospheric CO_2 contact with the experimental solution [38, 48, 52-53]. In addition, this approach had the advantage of controlling the solution saturation by controlling the pH. The researcher noted that it is mostly the CO_2 partial pressure that determines the precipitation formation [20-21, 42, 49-50, 58-60]. This phenomenon was used to control the CO_2 partial pressure of the surface of the tested solution to increase or decrease the liquid saturation [42, 49-50]. This enables the acceleration of the formation speed of the $CaCO_3$ precursors without modifying their dehydration speed [42, 49-50]. A predominant nucleation, of amorphous calcium carbonate, was detected at the lowest values of the used partial pressures while the higher partial pressure values result in heterogeneous nucleation, depending on the nature of the vessel material, and monohydrated calcium carbonate was detected [42, 52, 58, 61]. Gal et al, 1996, 2002 and Elfil and Roques, 2001, 2003 [20, 42, 47, 58] highlighted the fact that scaling conditions can be controlled by varying the CO_2 partial pressure in their experiments.

They declared that thermodynamic equilibrium conditions in addition to the water/air interface are of extreme importance. The real partial pressure of CO_2 in this interface region can be much lower than the equilibrium pressure because of an important loss of CO_2 at the interface. The pH values in this case will be higher than that calculated. This can be seen with the results of Boudeau and Canfield, 1993 [62] who found that closed system modelling agrees with actual findings with regards to pH and $CaCO_3$ saturation in seawater.

2.3. Materials

Accurate induction time measurements rely to a great extent on the accuracy of the method used in their measurement. Both the induction time measurement and the reactor design need to be able to perform under similar conditions encountered in the SWRO concentrate e.g. high ionic strength and no gas-liquid interface. Ease of sampling, temperature control and in addition the ability to measure frequently for a long period of time are advantages that may affect the induction time method's reliability. In order to satisfy these conditions, a series of induction time experiments were carried out to compare three widely-used methods, namely ICP, pH and conductivity measurements.

In addition, the effect of atmospheric CO_2 exchange due to mixing was evaluated by comparing the induction time measurements in an open and a closed system.

Inductively coupled plasma

The inductively coupled plasma (ICP) (Perkin Elmer Optima 3000) used in this research had a calcium detection limit of 10 µg/L. The instrument was used to measure the induction time by monitoring the change in calcium concentration due to the precipitation of $CaCO_3$ crystals. The ICP was connected to a diaphragm pump (Perkin Elmer) and fluid recirculation cooling system (Polyscience). The diaphragm pump is used for injection of samples to the plasma beam. The ICP was connected to a computer for data analysis using the ICP manufacturer software.

Conductivity meter

The conductivity meter (WTW) used in this research uses a solid non-refillable probe (Tetracon) with a measuring accuracy of 0.01 mS/cm in the range of total dissolved solids (TDS) lower than 7,000 mg/L and 0.1 mS/cm for higher TDS concentrates (>7,000 mg/L). The conductivity probe was mounted in a similar way to that of the pH on a stainless steel holder for the open reactor tests or fitted on the top cover of the glass reactor in the closed system. Conductivity calibration was carried out using a manufacturer's calibration solution twice per week. The conductivity measurements can be adjusted to be obtained automatically at a minimum of 60 second intervals.

pH meter

The induction time measurements using pH were performed by a highly sensitive pH meter (Eutech pH 6000) with an accuracy of 0.001 pH units. The pH meter is connected online for continuous measurement of pH over time. In the case of the open reactor, the pH meter was fitted on top using stainless steel holder while in the closed reactors, the reactor's top cover was equipped with custom-made holes for fitting the pH and conductivity probes (Figure 4). The pH measurements can be made online using the manufacturer's software or offline by using the instrument storage memory and the measuring intervals can be adjusted to as low as every 30 seconds.

Figure 4: The pH meter connected to the reactors

Reactors

Two types of reactors were used in the experiments; the first are open double jacketed glass reactors where and here the mixing was done using a magnetic stirrer. The second was a closed reactor equipped with a double paddled shaft mechanical stirrer. The mixing rate can be varied in both reactors from 0 to 1200 rpm. Either reactor can be filled mechanically using a diaphragm pump with an average filling speed of 4 L/min or manually using an elevated measuring glass flask as shown in Figure 5.

Figure 5: The glass reactor arrangements and connections

To avoid that during the initial stage of establishing the correct pH locally supersaturation will occur two separate solutions were prepared namely one with calcium chloride and sodium chloride and the other with hydrogen carbonate. The pH of the last solution was brought at such a level that after mixing these solutions in a ratio of 1:1 the intended pH would occur.

Theoretical calculations were conducted to predict the increase in concentration at points of solution addition resulting in the formation of local saturated zones with higher targeted solution saturation.

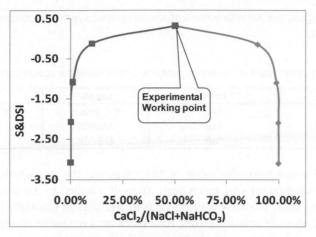

Figure 6: Schematic diagram for the experimental reactor

Calculation in terms of S&DSI, of the solution addition in the reactor revealed (Figure 6) that the maximum saturation reached for any specific point in the solution is only possible when complete mixing of the two reacting solutions has taken place. The maximum local saturation to be found at any specific point in the reactor is when the ratio of $(CaCl_2 + NaCl):HCO_3^-$ is 1:1. This saturation decreased rapidly by changing the ratios of these components with respect to each other. These results dismiss the hypothesis of formation of local saturation zones at the point of solution addition and confirm the role of thorough initial mixing to reach the desired experimental saturation.

After each experiment, a cleaning process employing 0.2 molar HNO_3 for 30 minutes with a flow of 0.15 L/min dissolves any formed crystals on the reactors walls or piping. The reactor is then flushed with dematerialized water for 15 minutes with a flow rate of 3 L/min before the start of the next experiment.

Synthetic seawater concentrate preparation

The ultra-pure water system is shown in Figure 7. Tap water is delivered as the raw water source where it passes through a series of treatment steps to decrease the organic and inorganic particles content in the feed water through filtration, softening, RO membrane, ion exchange and GAC filtration. The product water is finally disinfected using UV-light. The product water had a conductivity and total organic carbon (TOC) of 0.8 µS/cm and 3 µg/L, respectively. The TOC was measured using the TOC analyzer with a detection limit of 0.5 µg/L

The synthetic seawater concentrate used in the experiments was prepared in stages. Firstly, a $NaHCO_3$ solution was prepared by dissolving $NaHCO_3$ salt (Table 2) in ultrapure water. Secondly, a $CaCl_2.2H_2O$ solution was prepared by dissolving $CaCl_2.2H_2O$. Finally, NaCl salt was dissolved in the previously prepared $CaCl_2.2H_2O$

solution to adjust the salinity values of the prepared synthetic concentrate to the desired levels.

Table 2: Salt reagents used in the experimental synthetic seawater concentrate preparation

Reagent	Form	♀Supplier	Purity
NaCl	Salt	J.T. Baker	99.5-99.9 %
$CaCl_2.2H_2O$	Salt	MERCK	99.9
$NaHCO_3$	Salt	MERCK	99.9

To ensure the complete dissolving of the reagents, the preparation step uses dissolves the solution on a 1-L batch basis. The salt is added to the ultra-pure water in a measuring flask. The flask was then closed and shaken manually for 2 minutes after which 2 hours of solution mixing took place on a magnetic stirrer. Mixing was performed at an average speed of 400 rpm and at a room temperature of 20 °C.

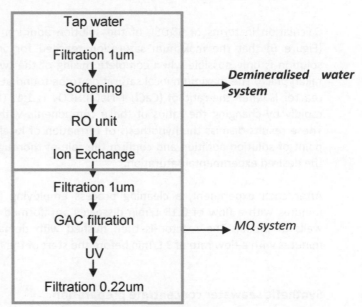

Figure 7: Laboratory ultra-pure water preparation system

The induction time experiments were initiated by adding the $NaHCO_3$ solution into the reactor followed by the NaOH solution for pH adjustment (if needed). Finally the $CaCl_2.2H_2O$ + NaCl solution was added at a rate of 0.2 L/min while maintaining a mixing speed of 150 rpm to ensure proper mixing and to prevent the formation of local saturation zones. The addition was performed through fine nozzles located 3 cm from the reactor's base to ensure proper distribution of the solution when added.

2.4.   Methods

Measurement of calcium with ICP

Two different methods are evaluated. In both methods samples are taken from the reactors and the formed calcium carbonate is removed by filtering through a membrane filter with pores of 0.2 um. In the first method the calcium concentration is measured in the solution at the start of the experiment and during the experiment after filtration (direct measurement).

In the second method the amount of precipitated calcium carbonate is measured by dissolving the calcium carbonate retained on the membrane filters (indirect measurement).

Salinity and accuracy of direct measurement of calcium in solution

Induction time can be detected by measuring the change of Ca^{2+} due to precipitation of $CaCO_3$. Measuring the induction time with this technique has already been explored and reported by many researchers [5, 31-36]. The ICP measuring procedure involved sampling of *e.g.* 5mL from the reacting solution by means of a plastic syringe every specific time interval (*e.g.* 5 min). The samples were filtered through 0.2 µm filter paper (Millipore) to remove any formed crystals followed by solution acidification to prevent any further reaction. The solution was then analyzed by ICP for Ca^{2+}.

For the purpose of sample measurement, a calibration curve was made (based on the expected amount of calcium to be measured) by preparing standard solutions of different concentrations of Ca^{2+} solutions (0, 0.5, 1, 2, 5, 10, 20, 29.6, 50, 267 mg Ca^{2+}/L) from a stock solution of 1000 mg/L calcium solution.. The ICP measurement technique depends on evaporating the measured sample instantaneously using a plasma beam and capturing the distinct measured ion wavelength and intensity. The intensity of the emitted waves is compared to that of the standard blank solutions for quantifying the amount of the studied ion. The instrument needs at least 2 mL of sample for each measurement where the sample passes through tiny hoses to the plasma beam for 45 seconds for plasma stabilization before measurement. Before and after each measurement, the feed hose is flushed by ultrapure water until the plasma beam restores its blue colour.

To explore the effect of salinity on the sensitivity of the ICP measurements, twelve solutions were prepared by dissolving (981.2 or 109.02 mg/L) of $CaCl_2.2H_2O$ containing (267 or 29.7 mg/L) of Ca^{2+} in ultrapure water. The TDS of the experimental solutions were varied using NaCl to reach a final solution concentration of (0, 1500, 10000, 30000, 60000, 100,000 mg/L) as shown in Table 3. These measurement were used to get and impression of the accuracy of this method, since the difference between the initial calcium concentration and the concentration during the experiments needs to be determined accurately.

Table 3: The solution composition for measuring the effect of salinity on induction time measurements (UPW is Ultra-pure water and PA is pure acid)

Sample	Ca^{2+} mg/L	NaCl mg/L	sample	Ca^{2+} mg/L	TDS mg/L
UPW	0	0	PA	0	0
1	267	0	7	29.7	0
2	267	1500	8	29.7	1500
3	267	10000	9	29.7	10000
4	267	30000	10	29.7	30000
5	267	60000	11	29.7	60000
6	267	100000	12	29.7	100000

Three samples were prepared using (i) ultrapure water,(ii) sample 1 and (iii) sample 7 containing 0; 267 and 26.7 mg/L of Ca^{2+}, respectively (Table3). The solutions were used making a calibration curve as shown in Figure 8. This curve was used afterwards by the instrument to determine the amount of Ca^{2+} in the tested samples. The measurement sequence was done in triplicate starting with the low Ca^{2+} value samples followed by the high Ca^{2+} values.

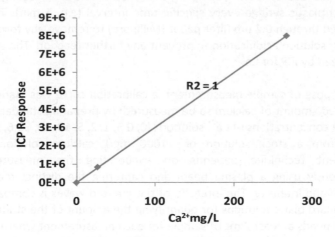

Figure 8: ICP calibration curve

Measurement of calcium carbonate retained on membrane filters

In order to investigate whether the measurement of calcium carbonate retained on membrane filters could give a higher accuracy and could eliminate the effect of salinity on the direct measurement of calcium in the solution after filtration. The induction time experiments used a 0.2 µm (Millipore) filter paper where a sample

(5mL) of the experimental solution was filtered every 5 minutes for the first 2 hours and every 30 minutes thereafter. The test was carried out on synthetic water of low ionic strength (I = 0.063 mole/L) as shown in Table 4. The saturation of the prepared solution according to the Stiff & Davis saturation index was 1.2 while the pH of the solution was 7.84.

To eliminate traces of dissolved Ca^{2+} retained inside the filter pores, as much as possible, the filter paper was made dry after filtering using a vacuum pump. Immediately after drying, the filter paper was soaked in 5 ml of 0.1 molar HCl (prepared from 10 molar HCl solution) to ensure a complete re-dissolution of the retained crystals. The acidified solution was then analyzed by ICP for calcium.

Errors in this new procedure may arise from three sources: firstly, the presence of calcium traces in the acid itself (the acid is diluted with ultrapure water); secondly, leaching of Ca^{2+} from the filter paper in the high acidic medium and thirdly, traces of dissolved Ca^{2+} trapped inside the filter pores after drying.

Therefore, three different blanks were made to measure these three possible experimental errors. The error due to the presence of calcium traces in the HCl solution was determined using a sample of the diluted HCl used in the crystal dissolving.

The error due to the possibility of calcium leaching from the filter paper was measured using a virgin filter which was soaked in the diluted HCl acid (the same procedure as carried out with the filter after filtering the sample).

Finally, the third error of entrapment of dissolved Ca^{2+} inside the filter pores was determined by filtering the $CaCl_2.2H_2O$ solution with the same concentration used in the experiments through the filter paper and dried in the same procedure used in the ICP measured filter sample. In such cases, 2.5 ml (half the sample taken from the reactor because the reagents are double the concentration inside the reactor) of the $CaCl_2.2H_2O$ reagent solution was filtered through a 0.2 µm and then dried with the same procedure described for measuring the induction time. The filter paper was then soaked in 5 ml of diluted HCl solution before being brought to the ICP for measurement.

Effect of salinity, stability and accuracy of conductivity and pH measurement

Induction time experiments were carried out on low saline and high saline concentrates as shown in Table 4. The investigation was done to determine the degree of accuracy and the stability of the conductivity and pH measurements with the increase in salinity.

For this purpose, two solutions were prepared; each contained the same amount of calcium and bicarbonate of 790 and 244 mg/L respectively. In the first solution the total dissolved solid (TDS) was 2526 mg/L and it had an ionic strength (I) of 0.063 mole/L. In the second solution the TDS and ionic strength were increased by the means of NaCl to values of 75908 mg/L and 1.32 mole/L, respectively. The induction time experiments were carried out in a closed reactor for nearly one day and the mixing speed was kept constant at 150 rpm.

Table 4: The effect of salinity on pH and conductivity measurements

Ions		Low ionic strength	High ionic strength
Temp.	°C	20	20
Calcium	mg/L	790	790
Sodium	mg/L	92	28987
Bicarbonate	mg/L	244	244
Chlorides	mg/L	1400	45957
TDS	mg/L	2526	75908
Ionic strength	mole/L	0.063	1.34

The pH and the conductivity measurements drop due to the precipitation of $CaCO_3$ can be quantified using the start and the end measured values as shown in equations 14-18. In the case of a pH drop the general equation is represented in equation 1 as follows:

$$Ca^{2+} + 2HCO_3^- \leftrightharpoons CaCO_3 + CO_2 \uparrow + H_2O \qquad 14.$$

At the experimental start ($t=i$)

$$pH_i = (pK_{a1})_i + log\frac{[HCO_3^-]_i}{[CO_2]_i} \qquad 15.$$

And after a time interval ($t=t$)

$$pH_t = (pK_{a1})_t + log\frac{[HCO_3^-]_t}{[CO_2]_t} \qquad 16.$$

As the TDS decrease due to precipitation is minor compared to the total TDS, the first acidity constant can be considered unchanged and therefore,

$$(pK_{a1})_i = (pK_{a1})_t \qquad 17.$$

Based on equation 8, the formation of $CaCO_3$ consumes 2 moles of HCO_3^- and produces 1 mole of CO_2 for each mole of formed $CaCO_3$. Therefore equation 3 can be re-written with respect to initial concentrations as represented in equation 12.

$$pH_t = (pK_{a1})_i + log\frac{[HCO_3^-]_i - 2X}{[CO_2]_i + X} \qquad 18.$$

Whereby solving equation 18 with regard to X, the amount of precipitated $CaCO_3$ can be determined.

The amount of X is also equivalent to the amount of calcium precipitated in mole/L from which the amount of precipitated $CaCO_3$ can be calculated. Theoretically, by using these equations a pH drop of 0.35 pH units represents 1 mg/L of precipitated $CaCO_3$

With regard to the conductivity, a similar concept can be adopted. The conductivity for water can be written as a function of the concentration of its ionic species (C_i) and their conductance factor (f_i) as shown in equation 19

$$EC_i = \sum C_i \cdot f_i \qquad\qquad 19.$$

Where after precipitation at time t, the equation can be written as shown in equation 20

$$EC_t = \sum C_t \cdot f_i \qquad\qquad 20.$$

For Ca^{2+} and HCO_3^-, the conductivity factor (f_i) is 2.6 & 0.715, respectively.

As only Ca^{2+} & HCO_3^- concentrations decreased in solution due to precipitation of $CaCO_3$, the difference in water conductivity can be written as in equation 21

$$EC_i - EC_t = \{((Ca^{2+})_i - ((Ca^{2+})_i - X))\} \cdot f_{Ca^{2+}} \\ + \{(HCO_3^-)_i - ((HCO_3^-)_i - 2X)\} f_{HCO_3^-} \qquad 21.$$

$$\therefore EC_i - EC_t = (X) f_{Ca^{2+}} + (2X) f_{HCO_3^-} \qquad\qquad 22.$$

When the decrease in EC_i - EC_t is known, the $CaCO_3$ precipitated can be calculated by solving equation 16 with respect to X. For example, the precipitation of 1 mg/L of $CaCO_3$ will result in a conductivity drop of 0.0025 mS/cm (2.5 μS/cm).

Effect of mixing on the induction time

Simulating the effect of the liquid-air interface effect on the induction time was explored using pH measurements at 20°C. The experiments were carried out using the same solution composition represented in Table 4, where the $NaHCO_3$ solution was added first while $CaCl_2.2H_2O$ solution was poured carefully while applying an initial high mixing speed (150 rpm) to prevent the formation of local saturation zones and to ensure a homogeneous solutions mixture within the beaker. The mixing speeds were set to the targeted mixing speeds (10, 150, 500 rpm) after completing the mixing of the two reagents.

The reliability of the induction time measurements requires on not initiating nucleation due to the formation of local saturation zones during the solution's addition with higher concentrations than the targeted final solution. For this purpose, theoretical sensitivity calculations were made to determine the maximum local saturation to be formed at the point of solution addition. This was simulated by assuming different levels of CaCl$_2$ / (NaCl+NaHCO$_3$) or NaHCO$_3$ / (NaCl+CaCl$_2$). The model calculations assumed a constant reactor volume of 1 litre where the $CaCl_2.H_2O$ +NaCl solution was varied from 0.01% to 99.99% of that volume. The rest of the volume was associated with NaHCO₃. To determine if changing the addition sequence might affect the model outcome, the calculations were done as well in reverse order where the NaHCO₃ was varied from 0.01% to 99.99% while the $CaCL_2.H_2O$ + NaCl solution was kept constant. The calculation simulations were carried out using the filling solution concentration (feed tanks concentrations) as these are double the targeted reactor concentrations and, when added on a 1:1 ratio, the end solution had the desired experimental ionic composition. In these calculations, the targeted mixture composition is shown in Table 5.

Table 5: Synthetic seawater concentrate ionic concentration inside the reactor at 20 °C

Ions	Concentration mg/L
Calcium	677.1
Bicarbonate	209.14
Sodium	20952
Chloride	33376
TDS	552151
Ionic strength	**0.96**

2.5. Results

Effect of salinity on the ICP measurements and accuracy.

Figures 8 and 9 and Table 6 show a pronounced effect of salinity of the response of the ICP in the Ca^{2+} measurements. The high salinity interferes with the calcium measurement where at salinities of seawater concentrates (nearly 60,000 mg/L) the measured values were 26% and 31% less than Ca^{2+} value in the solution, respectively. The change due to salinity can be immediately recognized even at very low salinities of 1500 mg/L (samples 2 and 8) as a decrease of 2.6 and 7.7% was recognized for Ca^{2+} levels of 267 and 29.7 mg/L, respectively. These results suggest the effect of salinity on the ICP measurements and can be linked to the analyzing technique used by the instrument. The ICP measurements are based on evaporating the tested solution and measure the spectra of the measured element. When there is a dominant element, Na^+ in our case, it appears that it affects the instrument's capability to capture the desired spectrum emission from the much less concentrated element.

In induction time experiments, this misleading decrease in calcium concentration will be considered as an immediate $CaCO_3$ precipitation (zero induction time) and conclusions will not represent the real induction time values. This instrumental measurement error is not constant, as shown in Figures 8 and 9, making it difficult to be predicted or calculated. In conclusion, applying ICP to the filtrate of high salinity concentrates in measuring induction time may lead to incorrect induction time values. As a consequence calibration lines have to be made for different salinities

Table 6: The effect of salinity on ICP measurements

Sample		Ca mg/l	NaCl mg/l	Reading		Average	% of change
				A	B		
No dilution	1	267	0	266.1	268.4	267.3	-0.1
	2	267	1500	261.0	259.8	260.4	2.6
	3	267	10000	240.7	237.7	239.2	10.5
	4	267	30000	218.2	216.6	217.4	18.7
	5	267	60000	197.9	194.8	196.4	26.5
	6	267	100000	168.5	172.1	170.3	36.3
Dilution 10 times	7	29.7	0	29.5	29.2	29.3	1
	8	29.7	1500	27.1	27.1	27.1	7.7
	9	29.7	10000	25.0	25.3	25.2	14.3
	10	29.7	30000	22.6	22.8	22.7	22.7
	11	29.7	60000	20.2	20.1	20.2	31.3
	12	29.7	100000	18.4	19.0	18.7	36.4

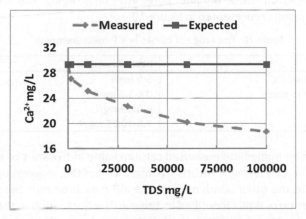

Figure 8: The measured calcium concentration by ICP (dashed line) with respect to TDS of the tested solution in comparison to the real concentration in the solution (solid line) for calcium content of 29.7 mg/L

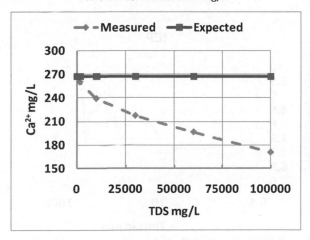

Figure 9: The measured calcium concentration by ICP (dashed line) with respect to TDS of the tested solution in comparison to the real concentration in the solution (solid line) for calcium content of 267 mg/L

Table 6 shows relative large differences between the duplicates (1.1 to 2.6 mg/L) which means that the method will be rather insensitive.

Accuracy of measurement of calcium carbonate retained on membrane filters

In the effort to improve the accuracy of the method and to eliminate the effect of salinity on the response of the ICP, samples were filtered through 0.2 µm filters. The filtrate was discarded and the filtered calcium carbonate precipitate was dissolved using HCl, and calcium concentration was measured on the ICP. Results represented in Table 7 showed that the inaccuracy in measurements using this approach varies between 7.5-10.25 mg/L of precipitated calcium (equivalent to 18.75 - 25.63 mg/L of precipitated $CaCO_3$). These high inaccuracies were mainly due to the retained calcium inside the filter pores after filter drying. This is shown in the induction time results represented in Figure 10.

Table 78: The range of blanks in ICP measurements

Type of Error	Min-Max mg/L
Entrapped dissolved Ca^{2+}	6-9 mg/L
Leaching from filter paper	0.6-1.2 mg/L
Acid dilution	0-0.05 mg/L
Total error	7.5-10.25 mg/L

Results showed an immediate measured calcium value at the start of the induction time experiment of 7.3 mg/L (nearly 18 mg/L of $CaCO_3$), suggesting immediate precipitation. On the other hand, most of the ICP measurements fell between the upper and lower error limits identified in Table 7. The fluctuation in the ICP calcium measurements in Figure 14 may suggest that the errors in measurements are dominating the calcium concentration measured, and that induction time cannot be practically identified.

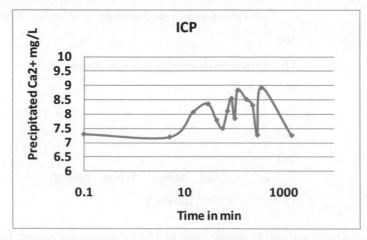

Figure 10: Induction time experimental result using calcium concentration vs. time using ICP for a low ionic strength water of I = 0.063

Efforts to further improving this method were not continued because of the successful application of a new highly sensitive and accurate pH meter suitable for saline solutions. See below.

The effect of salinity on the conductivity and the pH measurements

The measurement of induction time using conductivity and pH measurements showed a decrease in value with time indicating the precipitation of $CaCO_3$ (Figures 11 and 12).

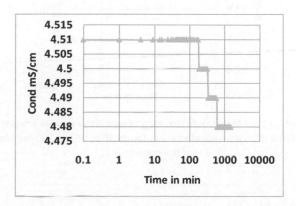

Figure 11: The induction time measurements using conductivity measurements for low saline concentrate; the first drop in conductivity was found after nearly 200 minutes from the start of the experiment

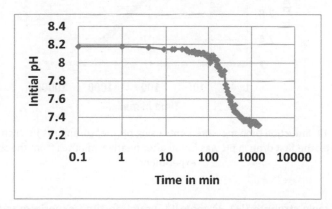

Figure 12: Induction time measurements using ΔpH measurements for low saline concentrate; the first drop in pH was found after nearly 10 minutes from the start of the experiment

In the low ionic strength (0.063 mole/L) experimental results, the induction time was clearly identified by both instruments (conductivity and pH). The first drop in conductivity of 0.01 mS/cm (10 µS/cm), which is equivalent to 2.5 mg/L of precipitated $CaCO_3$, was recorded after 200 minutes, while the pH measurements showed the first decrease of 0.01 pH units (10 times the detection limit and equivalent to 0.074 mg/L of precipitated $CaCO_3$) after nearly 10 minutes from the experiment start.

During 200 minutes the pH measurements dropped nearly 0.24 pH units (from 8.18 down to 7.94).

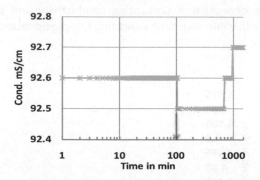

Figure 13: Induction time measurements using conductivity measurements for high saline concentrate; the first drop in conductivity was found after nearly 100 minutes from the start of the experiment

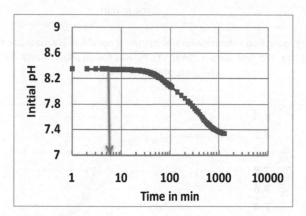

Figure 14: Induction time measurements using pH measurements for high saline concentrate; the first drop in pH was found after nearly 6 minutes from the start of the experiment

In the high ionic strength (I=1.34 mole/L) induction time experiments (Figure 13), the conductivity measurements showed the first drop (0.1 mS/cm) after 100 minutes before starting to increase again. As the conductivity meter sensitivity is a function in the initial solution conductance and equal to nearly 0.1%, the first recorded drop of 0.1 mS.cm was nearly 10 time higher than that found in low salinity water and in such case the drop is equivalent to nearly 25 mg/L of precipitated $CaCO_3$.

Meanwhile, the pH showed the first drop in its measurements (0.01 pH units drop) after only 6 minutes from the experimental start (Figure 14). The time of the first drop in conductivity was enough for the pH measurements to drop nearly 0.3 pH units. The results showed that the sensitivity of the pH measurements with respect to $CaCO_3$ induction measurements is nearly 15-20 times shorter than that provided by conductivity measurements. These results are supported by the theoretical effect of precipitation of calcium carbonate on conductivity and pH. Precipitation of 1 mg $CaCO_3$/L at the start of this experiment will theoretically result is a pH drop of 0.32 pH units for ionic strength of 1.34 mole/L, but 0.0025.mS/cm for the conductivity measurements. Since at high ionic strength, the accuracy of the pH meter is 0.001 and the accuracy of conductivity meter is 0.1 mS/cm, it is clear that from theoretical point of view the method making use of the new highly accurate pH meter is to be preferred as well.

The effect of carbon dioxide exchange in an open system

Exploring the effect of atmospheric CO_2 on the induction time measurements in an open system was done by measuring the pH drop at different levels of mixing. Results in Figure 15 show that mixing speed in an open reactor appears to affect the exchange of CO_2 with the solution, destabilizing the solution, and to facilitate $CaCO_3$ formation. The induction time measurements showed an inversely proportional relation between mixing speed applied and pH drop. At a mixing speed of 500 rpm the formation of crystals was instantaneously accompanied with a visual white cloud of crystals while at a lower mixing speed of 150 rpm it took nearly 2 minutes for the start of the pH drop and 10 minutes for the white cloud to start forming. The highest stability (100 minutes) before it precipitation commenced was obtained when a mixing speed of 10 rpm was applied. The 500 rpm experimental result showed a steeper slope (faster growth) and higher amount of precipitated $CaCO_3$ at the end of the experiment when compared to that of the 150 rpm experiment (1.25 times higher).

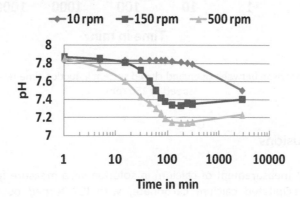

Figure 15: Induction time experiment in an open system with mixing speeds of 0, 150, and 300 rpm

A similar effect of mixing speed on the induction time has been discussed in literature for open systems. Drioli et al 2004 [46] confirmed such findings in unstirred solutions where the induction period was significantly longer and the reaction rates were substantially reduced. A possible reason for that may be the creation of surface turbulence and increase in the surface area exposed to direct contact with air (due to the formation of a vortex). This may help to increase mass transfer due to the increase in the Reynolds number [48]. The system stability under high saturation at 10 rpm mixing speed suggests that in an open system (CO_2 can exchange freely), CO_2 exchange with air considerably affects the induction time measurements. This result agreed with published results that induction time experiments are inversely proportional to the applied mixing speed [3, 20, 63], when taking into account that researchers in such cases ignored or marginalized the effect of CO_2 exchange with air in their results.

When a comparison was made between an open and a closed system at 150 rpm (Figure 16), the closed system showed a longer stability period before nucleation takes place (nearly 100 minutes) if compared to that of the open system (pH starts to decrease after just 2 minutes). The drop in pH measurements was slower in the closed system than that of the open system. The slow kinetics of formation confirms the effect of CO_2 exchange on the system stability.

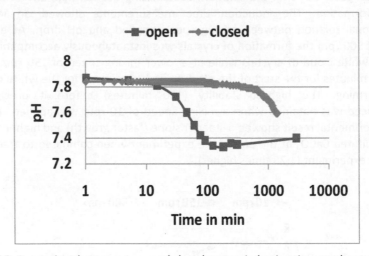

Figure 16: Comparison between open and closed system induction time results at a mixing speed of 150 rpm

2.6. Conclusions

- Direct measurement of calcium in solution, as a measure for the amount of precipitated calcium carbonate, with ICP turned out to be rather inaccurate to enable observing induction times. Moreover the response of the ICP was highly depended on salinity.

- Determination of calcium carbonate precipitated and retained on a membrane by dissolving in hydrochloric acid and measurement of calcium ICP suffered from high blanks. The main source of these high blanks is the remaining calcium, origination from the solution, in the membrane filters.

- Conductivity measurements were inaccurate as well, since the change in conductivity due to precipitation of calcium carbonate was too low to be detected accurately with this method.

- Measurements with a new (on the market) highly accurate and stable pH meter were far the most accurate and sensitive. As a consequence this method has been selected for determination of the induction times in this study.

- Experiments with open reactor showed a large effect of exchange of carbon dioxide with the atmosphere. Consequently closed reactors are apllied in this study.

2.7. List of Symbols

A	Function of Pre-exponential factor ($s^{-1}m^{-3}$)
A_f	Free area for precipitation at given particle size (m^2)
Alk	Alkalinity of solution (mole/L)
B	Constant expressed ($L^{3/2}mole^{-1}/m^{-1}$)
C	Concentration
EC	Conductivity ($\mu S/cm$)
f	Conductance factor ($\mu S/cm$ per mg/L)
I	The ionic strength (mole/L)
IAP	Ionic activity product ($mole^2/L^2$)
J	Nucleation rate (nuclei/min/cm^3)
K_a	Area geometric factor
K^T_{a2}	Temperature corrected second acidity constant (mole/L)
k_b	Boltzmann constant (J/K)
K_{so}	Solubility at standard conditions ($mole^2/L^2$)
K_{sp}	Solubility product ($mole^2/L^2$)
K	Graphical obtained constant in S&DSI calculations
K_v	Volume geometric factor
Log	Log_{10}
N	Number of molecules
P	Pressure (psi)
pAlk	Negative the logarithm of alkalinity divided by its dimensions
pCa	Negative the logarithm of calcium divided by its dimensions
pH	Concentrate pH
pH_s	Equilibrium pH

S_a	Supersaturation ratio
SI	Supersaturation Index
r	Nuclei radius (m)
r_{crit}	Critical nuclei radius (m)
T	Absolute temperature (Kelvin)
t_{ind}	Induction time in minutes unless mentioned otherwise (min)
υ	Number of ions into which a molecule dissociates

Greek letters

$f(\theta)$	Factor differentiating heterogeneous and homogenous nucleation
β	Geometric factor
ϑ	Molecular volume (cm^3/mole)
$\gamma+$	Cation activity coefficient
γ_-	Anion activity coefficient
γ_s	Surface energy (J/m^2)
Ω	Pre-exponential factor in the nucleation rate equation ($s^{-1}m^{-3}$)
ϕ	Pre-exponential factor in the nucleation rate equation

2.8. References

1.   ASTM, Calculation and Adjustment of the Stiff and Davis Stability Index for Reverse Osmosis. 2001, ASTM International: West Conshohocken, PA, United States.

2.   ASTM, Calculation and Adjustment of the Langelier Saturation Index for Reverse Osmosis. 2003, ASTM International: West Conshohocken, PA, United States.

3.   Sohnel, O. and J. Garside, Precipitation basis principals and industrial applications. 1992, Oxford: Butterworth-Heinemann.

4.   Stiff, H.A. and L.E. Davis, A method for predicting the tendency of oil field waters to deposit calcium carbonate. Petroleum transactions, 1952. 195: p. 213-216.

5.   Boerlage, S., Scaling and particulate fouling in membrane filtration system, in Sanitary Engineering. 2002, IHE: Delft.

6.   Darton, E., Membrane chemical research: centuries apart. Desalination, 2000. 132: p. 121-131.

7.   Sohnel, O. and J.W. Mullin, Precipitation of calcium carbonate. Journal of Crystal Growth, 1982. 60: p. 239-250.

8.   Sohnel, O. and J.W. Mullin, Influence of mixing on batch precipitation Crystal Research and Technology 1987. 22(10): p. 1235 - 1240.

9.   Hasson, D., et al., Detection of fouling on RO modules by residence time distribution analyses. Desalination, 2007. 204(1-3): p. 132-144.

10.   Van Gauwbergen, D. and J. Baeyens, Macroscopic fluid flow conditions in spiral-wound membrane elements. Desalination, 1997. 110: p. 287-299.

11.   Vrouwenvelder, J.S., Biofouling of spiral wound membrane system. 2009, Delft University of Technology: Delft.

12.   Boerlage, S., et al., Stable barium sulphate supersaturation in reverse osmosis. J. Mem. Sci., 2000. 179: p. 53-68.

13.   Abdel-Aal, E., M. Rashad, and H. El-Shall, Crystallization of calcium sulfate dihydrate at different supersaturation ratios and different free sulfate concentrations. Cryst. Res. Technol., 2004. 39: p. 313–321.

14.   Taguchi, K., J. Garsid, and N.S. Tavare, Nucleation and growth kinetics of barium sulphate in batch precipitation. Journal of Crystal Growth 1996. 163 p. 318-328.

15.   Symeopoulos, B.D., Spontaneous Precipitation of Barium Sulfate in Aqueous Solution. J. CHEM. SOC. FARADAY TRANS., 1992. 88(20): p. 3063-3066.

16.   Verdoes, D., D. Kashchiev, and G.M.v. Rosmalen, Determination of nucleation and growth rates from induction times in seeded and unseeded precipitation of calcium carbonate. Journal of Crystal Growth, 1992. 118 p. 401-413.

17.   Malollari, I.X., P.G. Klepetsanis, and P.G. Koutsoukos, Precipitation of strontium sulfate in aqueous solutions at 25°C. Journal of Crystal Growth, 1995. 155 p. 240-246.

18.   Sergei, V., O.S. Pokrovsky, and V.S. Savenko, Unseeded precipitation of calcium and magnesium phosphates from modified seawater solutions. Journal of Crystal Growth, 1999. 205: p. 354-360.

19.   Golubev, S.V., O.S. Pokrovsky, and V.S. Savenko, Unseeded precipitation of calcium and magnesium phosphates from modified seawater solutions. Journal of Crystal Growth, 1999. 205: p. 354-360.

20.   Gal, J., Y. Fovet, and N. Gache, Mechanisms of scale formation and carbon dioxide partial pressure influence.Part II. Application in the study of mineral waters of reference. Water Research, 2002. 36: p. 764–773.

21.   Ledion, J., B. Francois, and J. Vienne, Characterization of scaling power of some water by fast controlled precipitation. J. Eur. Hydr., 1997. 28(1): p. 15-35.

22.   Roelands, C.P.M., Polymorphism in Precipitation Processes. 2005, TU Delft: Delft.

23.   Prisciandaro, M., A. Lancia, and D. Musmarra, Calcium Sulfate Dihydrate Nucleation in the Presence of Calcium and Sodium Chloride Salts. Eng. Chem. Res.I 2001. 40: p. 2335-2339.

24.   El-Shal, H., et al., A study of primary nucleation of calcium oxalate monohydrate:Effect of supersaturation. Cryst. Res. Technol., 2004. 39(3): p. 214 – 221.

25.   Turek, M. and P. Dydo, Electrodialysis reversal of calcium sulphate and calcium carbonate supersaturated solution. Desalination, 2003. 158: p. 91-94.

26.   Lisitsin, D., D. Hasson, and R. Semiat, Critical flux detection in a silica scaling RO system. Desalination, 2005. 186(1-3): p. 311-318.

27.   Turek, M., P. Dydo, and J. Waś, Electrodialysis reversal in high CaSO$_4$ supersaturation mode. Desalination 2006. 198 p. 288-294.
28.   Hasson, D., A. Drak, and R. Semiat, Inception of CaSO$_4$ scaling on RO membranes at various water recovery levels. Desalination, 2001. 139(1-3): p. 73-81.
29.   Hasson, D., A. Drak, and R. Semiat, Induction times induced in an RO system by antiscalants delaying CaSO$_4$ precipitation. Desalination, 2003. 157(1-3): p. 193-207.
30.   Clarkson, J.R., T.J. Price, and C.J. Adams, Role of Metastable Phases in the Spontaneous Precipitation of Calcium Carbonate. J. chem. soc. faraday trans, 1992. 88(2): p. 243-249
31.   Tarasevich, B.J., et al., The nucleation and growth of calcium phosphate by amelogenin. Journal of Crystal Growth 2007. 304: p. 407-415.
32.   Wang, Y., Composite fouling of calcium sulphate and calcium carbonate in a dynamic seawater reverse osmosis unit. 2005, University of New South Wales: Sydney.
33.   Butt, F.H., F. Rahman, and U. Baduruthamal, Evaluation of SHMP and advanced scale inhibitors for control of CaSO$_4$, SrSO$_4$, and CaCO$_3$ scales in RO desalination Desalination, 1997. 109: p. 323-332.
34.   Mustafa, G.M., The study of pretreatment options for composite fouling of reverse osmosis membranes used in water treatment and production, in Chemical science and Engineering. 2007, University of New South Wales.
35.   Yu, H., The mechanism of composite fouling in Australian sugar mill evaporators by calcium oxalate and amorphous silica, in Chemical Engineering and Industrial chemistry. 2003, University of New South Wales: Sydney.
36.   Sheikholeslami, R. and H.W.K. Ong, Kinetics and thermodynamics of calcium carbonate and calcium sulfate at salinities up to 1.5 M. Desalination, 2003. 157(1-3): p. 217-234.
37.   Corre, K.S.L., et al., Impact of calcium on struvitecrystal size, shapeand purity. Journal of Crystal Growth, 2005. 283: p. 514-522.
38.   Fathi, A., et al., Effect of a magnetic water treatment on homogeneous and heterogeneous precipitation of calcium carbonate. Water Research 2006. 40: p. 1941 - 1950.
39.   Doğan, Ö., E. Akyol, and M. Öner, Polyelectrolytes inhibition effect on crystallization of gypsum. Cryst. Res. Technol., 2004. 39(12): p. 1108 - 1114.
40.   Hu, H., et al., A spectrophotometer-based method for crystallizationinduction time period measurement. Journal of Crystal Growth, 2001. 232 p. 86-92.
41.   Alimi, F., H. Elfil, and A. Gadrib, Kinetics of the precipitation of calcium sulfate dihydrate in a desalination unit. Desalination, 2003. 157(9 -16).
42.   Elfil, H. and H. Roques, Role of hydrate phases of calcium carbonate on the scaling phenomenon. Desalination, 2001. 137: p. 177-186.
43.   Hannachi, A., et al. A new index for scaling assessment. in IDA World Congress-Maspalomas. 2007. Gran Canaria.

44.   Chen, T., A. Neville, and M. Yuan, Assessing the effect of Mg^{2+} on $CaCO_3$ scale formation–bulk precipitation and surface deposition. Journal of Crystal Growth, 2005. 275: p. 1341-1347.

45.   Xie, A.-J., et al., Influence of calcium binding proteins on the precipitation of calcium carbonate: A kinetic and morphologic study. Cryst. Res. Technol. , 2006. 41(12): p. 1214 - 1218.

46.   Koutsoukoas, P. and C. Kontoyannis, Precipitation of calcium carbonate in aqueous solutions. J. Chem. Soc., Faraday Trans.I, 1984. 80: p. 1181 1192.

47.   Gal, J., et al., Calcium carbonate solubility: a reappraisal of scale formation and inhibition. Talanta, 1996. 43: p. 1497-1509.

48.   Drioli, E., et al., Integrated system for recovery of CaCO3, NaCl and $MgSO_4 \cdot 7H_2O$ from nanofiltration retentate. Journal of Membrane Science, 2004. 239: p. 27-38.

49.   Morales, J.G., J.T. Burgues, and R.R. Clemente, Nucleation of calcium carbonate at different initial pH conditions. Journal of Crystal Growth, 1996. 166: p. 331-338.

50.   Morales, J.G., et al., Precipitation of calcium carbonate from solutions with varying Ca^{2+}/carbonate ratios. Journal of Crystal Growth 1996. 166: p. 1020-1026.

51.   S. Bagus, P., Y.S. Lee, and K.S. Pitzer, Effects of relativity and of the lanthanide contraction on the atoms from hafnium to bismuth. Chemical Physics Letters, 1975. 33(3): p. 408-411.

52.   Elfil, H. and H. Roques, Prediction of the limit of the metastable zone in the $CaCO_3$-CO_2-H_2O system. AICHE journal, 2004. 50(8): p. 1908-1916.

53.   Gabrielli, C., et al., Nucleation and growth of calcium carbonate by an electrochemical scaling process.Journal of Crystal Growth, 1999. 200: p. 236-250.

54.   Sheikholeslami, R., Mixed salts--scaling limits and propensity. Desalination, 2003. 154(2): p. 117-127.

55.   Butler, J., Carbon Dioxide equilibria and their applications. second ed. 1982, California: Addison-Wesley.

56.   Butler, J., Ionic Equilibrium: Solubility and pH calculations. 1998, New York: John Wiley & Sons.

57.   Tlili, M.M., A.S. Manzola, and M. Ben Amor, Optimization of the preliminary treatment in a desalination plant by reverse osmosis. Desalination, 2003. 156: p. 69-78.

58.   Elfil, H. and H. Roques, Kinetics of the precipitation of calcium sulfate dihydrate in a desalination unit. Desalination 2003. 157: p. 9-16.

59.   Gordon, L.I. and L.B. Jones, The effect of temperature on carbon dioxide partial pressures in seawater. Marine Chemistry, 1973. 1: p. 317-322.

60.   Roques, H.E.a.H., Role of hydrate phases of calcium carbonate on the scaling phenomenon. Desalination 2001. 137: p. 177-186.

61.   Elfil, H. and H. Roques, Prediction of limit of metastable zone in the $CaCO_3$-CO_2-H_2O system. AIChE journal, 2004. 50(8): p. 1908-1916.

62.   Boudreau, B.P. and D.E. Canfield, A comparison of closed- and open-system models for porewater pH and calcite-saturation state. Geochimica et Cosmochimica Acta, 1993. 57: p. 317-334.

63.   Golubev, S.V., O.S. Pokrovsky, and V.S. Savenko, Homogeneous precipitation of magnesium phosphates from seawater solutions. Journal of Crystal Growth, 2001. 223: p. 550-556.

Chapter 3

The effect of particles on the induction time of calcium carbonate in synthetic SWRO concentrate

Published as:
Waly, T.; Munoz, R.; Kennedy, M.D.; Witkamp, G.J.; Amy, G. and Schippers, J.C. Role of particles on calcium carbonate scaling of SWRO systems. Desalination and Water Treatment, 2010. 18 p103-111
Waly, T.; Munoz, R.; Kennedy, M.D.; Witkamp, G.J.; Amy, G. and Schippers, J.C. Role of particles on Calcium carbonate scaling of SWRO systems. In proceeding of the EDS (Ed.) desalination for the environment conference. 2009. Baden-Baden, Germany

3.1. Abstract

Particles, which are naturally present in seawater, have been reported to affect the formation of $CaCO_3$ crystals. In this study the effect of foreign particles on induction time was explored. Induction time was monitored using a highly sensitive pH meter mounted in an air-tight glass reactor. Experimental work was performed with synthetic seawater prepared using ultra-pure water with a composition equivalent to 50% recovery SWRO concentrate (based on the composition of seawater from the Gulf of Oman). The prepared synthetic solutions were passed through filters which had different pore sizes, namely 0.22 μm; 0.1μm and 100kD ≠0.03 μm). In addition, glass beads of diameter 20-30 nm were added in volumes of (1; 0.5 and 0.1 ml) to synthetic seawater to yield an increase in glass surface area in contact with experimental synthetic water by 123%, 16% and 1% compared to the total reactor glass surface area. Different mixing speeds were applied, namely 10; 50; 150 and 300 rpm, to investigate the effect of mixing on the experimental results. Results showed that filtering synthetic seawater and applying different mixing speeds had a negligible effect on the experimental induction time results. On the contrary, the addition of glass beads shortened the induction time substantially from 30% till 100% (immediate precipitation), depending on the amount of beads added.

Keywords: Induction time, membrane, calcium carbonate, glass beads, filtration

3.2. Background

Reducing the consumption of chemicals in seawater reverse osmosis (SWRO) plants is an important goal because of the positive impact on the environment and the reduction in operational costs. Acid/antiscalant dosing to avoid calcium carbonate scale is responsible for a major part of the chemical consumption in SWRO. Reducing chemical dosing can be achieved by determining the actual scaling limits of calcium carbonate in SWRO, and operating close to these limits. This can be achieved by investigating the factors affecting the nucleation mechanism of $CaCO_3$ in seawater concentrates.

Particles generally exist in natural aquatic environments and seawater is no exception to this rule. Indeed, these particles might be different in terms of their nature, size and concentration, but generally they can be classified as: settleable solids (>100 um), supra-colloidal solids (1 um to 100 um), colloidal solids (0.001 um to 1um) and dissolved solids (<0.001 um) [1]. The most common inorganic colloids in natural waters are aluminium silicate clays and colloids of iron, aluminium and silica. Organic deposits, together with iron oxide, silica and aluminium, represent 70% of the deposits detected in membrane autopsies throughout the world [2-3]. Park et al. measured in natural seawater around 10^7 to 10^{10} particle/L with an average diameter of 128 nm [4] which results in a surface area of $5*10^{-4}$ m^2/L.

The effect of the presence of particles on nucleation and subsequent growth has been addressed by many researchers [5-12], and it has been claimed that active impurities, e.g. macro-molecules, organic compounds and metallic ions, may have growth suppression effects on precipitated salts depending on the size, shape, orientation, and molecular size of the impurity [5, 11-12]. On the contrary, it has also been reported that the presence of particles may affect the mechanism of nucleation of a particular scalant due to the generation of extra nucleation sites associated with the settling of these particles [10, 13]. Thompson, 2003 [14] claimed that impurities in the solution induce their effects by interacting with crystal faces during growth; some impurities can completely suppress growth, some enhance growth, while others act selectively or to varying degrees on each crystal face, consequently modifying the crystal morphology [14]. On the contrary, Snoeyink and Jenkins, 1980 [15] demonstrated a negative effect of particles on slightly supersaturated solutions with respect to $CaCO_3$, where these solutions showed an infinite degree of stability until fine foreign particles were added. The resultant solution performance in such cases was very similar to what is found if supersaturation is increased [15].

Scaling is considered a continuous potential hazard in the design and operation of SWRO systems. Calcium carbonate, as the most abundant scalant in SWRO, is commonly assumed to precipitate immediately when its solubility is exceeded. The supersaturation of a solution with regard to calcium carbonate can be calculated using different indices. The most commonly used indices are the Stiff & Davis stability index (S&DSI); Saturation Index (SI) and Saturation ratio (S_a) [16-23]. The S&DSI approach is based on correcting the values of the solubility product and second acidity constant for salinity and temperature based on experimental data [22]

$$S\&DSI = pH - pH_s \qquad\qquad\qquad\qquad\qquad\qquad 1.$$

$$pH_s = pCa + pAlk + k \qquad\qquad\qquad\qquad\qquad 2.$$

And

$$k = pK_{a2}^{TNaCl} - pK_{sp} \qquad\qquad\qquad\qquad\qquad 3.$$

When it comes to practice, supersaturated solutions normally exhibit a period of stability after which precipitation takes place [21]. This period of stability is defined as the induction time, and is the time elapsed between the creation of supersaturation and the first appearance of a detectable new phase [21]. Induction time is commonly measured by monitoring the change of water parameters over time e.g. pH, conductivity, turbidity or the concentration of specific crystal ions [24].

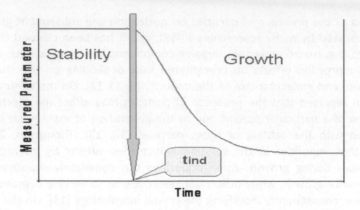

Figure 1: Schematic diagram showing the experimental detection of induction time

Experimentally, it is very difficult to determine the formation of the first nuclei, and consequently, a part of the induction time may also include growth to a detectable size.

The induction time was reported to be affected by the stirring speed applied during the reaction, and was reported to decrease with increasing solution supersaturation and to increase as the applied mixing speed increased (300 to 700 rpm) [21, 25-26]. This relation may be attributed to the method of measuring induction time and whether the nucleation mechanism was homogenous or heterogeneous [27-28], as homogenous nucleation was reported not to be affected by changing the mixing speed [29]. A similar effect was reported by Trainer, 1981 and Wiechers et al, 1981 on their crystal growth experiments where results showed that stirring speeds of 300 to 800 rpm had no effect on the growth mechanism [27-28].

For the purposes of this research it is assumed that the nucleation time is much greater than the time required for growth of crystal nuclei to a detectable size [30]. Thus, induction time can be assumed to be inversely proportional to the rate of nucleation, t_{ind} α J^{-1} [21]

$$t_{ind} \propto J^{-1} \hspace{4cm} 4.$$

Where the steady state nucleation rate can be expressed in

$$J == \Omega_{het} exp\left(\frac{-\beta\vartheta^2\gamma_s^3 f(\theta)}{k^3 T^3 v^2 Ln^2 S_a}\right) \hspace{3cm} 5.$$

In equation 5 an arbitrary factor (ƒ(θ)) determines the nucleation mechanism. Homogeneous nucleation is recognized by ƒ(θ) = 1 while for heterogeneous nucleation ƒ(θ) < 1. Values as low as 0.01 were reported [21]

Based on equations 4 & 5

$$logt_{ind} = \frac{B}{(T)^3 . log^2 S_a} - A$$
6.

Based on nucleation theory, homogenous and heterogeneous nucleation mechanisms are basically differentiated by the description of nucleus formation. Nucleation is referred to as homogeneous nucleation when the formation of a solid phase is not influenced by the presence of any solid phase, while in heterogeneous nucleation, the formation of new solid phase particles is catalyzed by the presence of a foreign solid phase. When nucleation occurs, the growth of a solid phase is then initiated by the presence of the solid phase already formed [21]. The solid phases inside a SWRO module can be particles present in the feed water or the membrane surface itself. In literature, homogenous and heterogeneous nucleation are differentiated by the slope of the relation between Log t_{ind} and Log S_a^{-2} shown in equation 6 [6, 11-12, 31]. The change in slope is related to the geometrical shape β in equation 5 which represents different crystallization habits at different ranges of supersaturation. For $CaCO_3$, this boundary saturation ratio (S_a) at which β changes range from 20 [12] up to 100 [31].

3.3. Materials

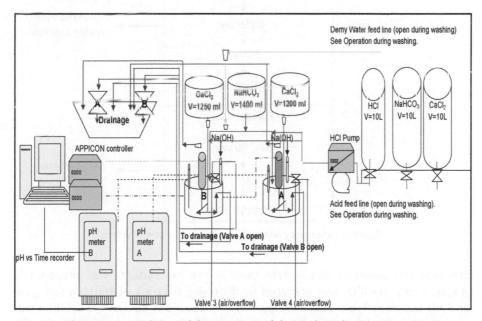

Figure 2: Schematic general view of the system used during the induction time experiments

The experimental setup shown in Figure 2 was built to measure the induction time using a highly sensitive on-line pH meter (Eutech pH 6000) with an accuracy of 0.001 pH units. The pH meter is connected on-line for continuous measurement of pH over time. The pH probe was fitted in the top of the air-tight double-jacketed glass reactor with a volume of 3 litres (Applikon), and equipped with a double-paddled shaft mechanical stirrer. The mixing rate can be varied from 0 to 1200 rpm using an electronic controller (Applikon) linked to the mixing motor. The reactor can be filled either mechanically using a diaphragm pump with average filling speed of 4 L/min or manually. After each experiment the reactor was cleaned for 30 minutes, employing 0.2 molar HCl or HNO_3 to dissolve any crystals formed with a flow of 0.15 L/min, and then flushed with demineralised water for 15 minutes with a flow rate of 3 L/min before the next experiment.

Synthetic seawater concentrate preparation

The ultra-pure water system is shown in Figure 3. Tap water is delivered as the raw water source where it passes through a series of treatment steps to decrease the organic and inorganic particle content in the feed water. The product water had a conductivity and total organic carbon (TOC) 0.8 µS/cm and 3 µg/L, respectively. The TOC was measured using a TOC analyzer with a detection limit of 0.5 µg/L

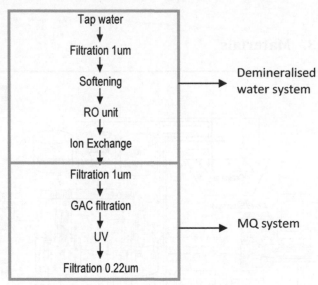

Figure 3: Laboratory ultra-pure water preparation system

The synthetic seawater concentrate used in the experiments was prepared in stages. Firstly, $NaHCO_3$ was prepared by dissolving 0.81 g/L of $NaHCO_3$ salt (pro-analysis shown in Table 1) in ultra-pure water. Secondly, $CaCl_2.2H_2O$ solution was prepared by dissolving 5.25g/L of $CaCl_2.2H_2O$. Finally, 17.93 g/L of NaCl salt was dissolved in the previously prepared $CaCl_2.2H_2O$ solution to adjust the ionic strength value (1.62 mole/L) of the prepared synthetic concentrate to compensate for the absence of other ions present in real seawater concentrates.

To ensure the complete dissolution of reagents, the salt was added to the ultra-pure water in a measuring flask. The flask was then closed and shaken manually for

2 minutes after which 2 hours of solution mixing took place on a magnetic stirrer. Mixing was performed at an average speed of 400 rpm and at a room temperature of 20 °C.

Table 1: Salt reagents used in the experimental synthetic seawater concentrate preparation

Reagent	Form	Supplier	Purity
NaCl	Salt	J.T.Baker	99.5-99.9 %
$CaCl_2.2H_2O$	Salt	MERCK	99.9
$NaHCO_3$	Salt	MERCK	99.9

The induction time experiments were initiated by adding the $NaHCO_3$ solution into the reactor followed by the NaOH solution for pH correction. Finally the $CaCl_2.2H_2O$ + NaCl solution was added with a rate of 0.2 L/min while maintaining a mixing speed of 150 rpm to ensure proper mixing and to prevent the formation of local saturation zones. The addition was performed through fine nozzles located 3 cm from the reactor's base to ensure proper distribution of the solution when added. The two reacting solutions were added on a 1:1 volume basis. The resultant synthetic solution is equivalent to 50% recovery SWRO concentrate using feed water from the Gulf of Oman (Table 2).

Table 2: The ionic composition of the experimental synthetic seawater concentrates equivalent to 50% recovery using Gulf of Oman water.

Ions		Experimental Solution
Calcium	mg/L	948
Bicarbonate	mg/L	293
Sodium	mg/L	35,382
Chloride	mg/L	56,071
Ionic Strength	Mole/L	1.61

3.4. Methods

The procedure used to study the effect of particles on $CaCO_3$ nucleation comprised three different experiments:

1. the effect of particles in synthetic concentrate by filtration.
2. addition of foreign particles
3. effect of mixing on the experimental results.

The effect of particles in synthetic seawater on the induction time

Prior to the start of the induction time experiment the prepared solutions were first filtered separately through either a 0.2μm; 01μm or 100k≈D₂0.03μm), polyethersulphone (PES) membrane (Millipore) with a diameter of 76 mm. Before use, the filters were flushed and then soaked for 1 day in a clean plastic cup, which was completely filled with ultra-pure water and sealed. Filtration was performed under pressure using an Amicon stainless steel stirred cell at a pressure of 3 bars.

The pH was monitored over a period of at least 1000 minutes and induction time was defined as the time corresponding to a pH drop of 0.03 pH units. This decrease in pH was equivalent to 0.3 mg/L of precipitated $CaCO_3$. The pH of the synthetic concentrate was varied between 7.8 and 8.6 to simulate different levels of supersaturation using 0.2 M NaOH prepared from pure salt (99.99% MERCK) that was dissolved in ultra-pure water. The Stiff and Davis saturation index was calculated for each experiment, and the S&DSI vs. induction time was plotted independently for each test.

The effect of particle addition

The number of particles in the synthetic seawater water was increased by adding inert glass nano-particles with a diameter ranging from 20-30 nm. The particles were introduced into the reactor as a suspension (48% water content supplied by BaseClear). The particles were injected in 3 volumes of either 0.1; 0.5 or 1 ml of suspension producing a particle content of $6.4*10^{-10}$; $8*10^{12}$ or $6.4*10^{13}$ particle/L, respectively. These glass particles produced an extra surface area for nucleation equal to $4.2*10^{-5}$; 0.005 or 0.042 m^2/L. Taking into consideration the surface area of the glass reactor walls, mixer shaft, mixing paddle and reactor (metal) cover in contact with the solution (as shown in Table 3), the increase in surface area is equivalent to 0.09%; 11.6% and 92.6% for the 0.1; 0.5 and 1 mL doses, respectively.

Table 3: Surface area after the addition of 20-30 nm glass beads in the experimental synthetic seawater concentrates.

	Units	Particle addition to synthetic seawater		
		1 ml	0.5 ml	0.1 ml
No. of particles	Particle/L	6.4E+13	8E+12	6.4E+10
Area/L	m^2/L	0.042	0.005	4.19E-05
Area reactor/L	m^2/L		0.034	
Area other parts/L	m^2/L		0.0112	
Area particles/ Area glass	%	123.1	15.4	0.12
Area particles/ Area total	%	92.6	11.6	0.09

The glass particles were added after adding the $NaHCO_3$ solution and mixing was maintained at 150 rpm to insure a proper distribution in the solution. The addition of NaOH and $CaCl_2.2H_2O$+NaCl solutions followed the same procedure described in the synthetic water preparation. Three different levels of saturation were tested, namely, S&DSI = 0.77; 0.62 and 0.55 corresponding to an initial pH of 8.15; 8.0 and 7.93

Mixing effect

The mixing rate was varied to determine the effect of diffusion on induction time. Three different mixing speeds, namely 10; 50 and 300 rpm, were tested and compared to the results obtained at 150 rpm.

3.5. Results and discussion

The effect of particles in synthetic seawater on the induction time

Figure 4 shows a sample pH measurement over time for synthetic seawater concentrates pre-filtered through 100kDa (≈0.03 μm) filters. The initial S&DSI was 0.85, and the induction time based on the developed method was estimated to be 25 minutes.

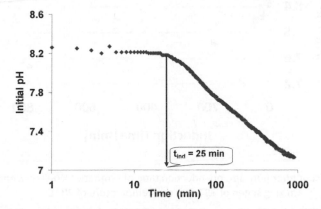

Figure 4: Initial pH vs. Time (minutes) for synthetic SWRO concentrates (recovery 50%) at mixing speed of 150 rpm and temperature of 20 °C

Table 4: Synthetic concentrate pH and corresponding Stiff and Davis saturation Index value

Unfiltered			0.2µm filtered			0.1µm filtered			100KDa (≈0.03 µm) filtered		
pH	t_{ind} min	S&D SI	pH	t_{ind} min	S&D SI	pH	t_{ind} min	S&D SI	pH	t_{ind} min	S&DSI
7.91	101	0.47	7.83	181	0.39	7.51	778	0.07	7.56	554	0.12
8.20	55	0.76	8.20	44	0.76	7.92	114	0.48	7.82	168	0.38
8.23	41	0.79	8.25	28	0.81	8.09	49	0.65	8.09	60	0.65
8.27	20	0.83	8.31	21	0.87	8.25	68	0.81	8.25	24	0.81
8.34	19	0.90	8.41	20	0.97	8.34	34	0.90	8.52	18	1.08
8.52	11	1.08	8.43	9	0.99	8.41	22	0.97	8.54	10	1.10
8.64	4	1.20				8.53	13	1.09	8.60	13	1.16
						8.62	7	1.18			
						8.67	7	1.23			

Experiments were performed with synthetic seawater filtered through 0.2 μm; 0.1 μm and 100 kDa (0.03 μm) filters. The results presented in Table 4 and Figure 5 showed that there is no pronounced effect of filtration on the induction time. This may be attributed to the fact that the ultra-pure water and pro-analysis chemicals used in the preparation of our synthetic seawater resulted in more or less particle-free synthetic concentrates or that particles greater than 100 kDa (0.03 μm) have a minor effect on the measured induction time. The logarithmic relation between the initial pH and the corresponding induction time agreed with what was previously reported by Waly et al, 2009 [23]. The correlation factor (R^2) for the graph is 0.93; 0.94; 0.92 and 0.94 for unfiltered; 0.2μm; 0.1μm and 100 kDa (0.03 μm) filters respectively.

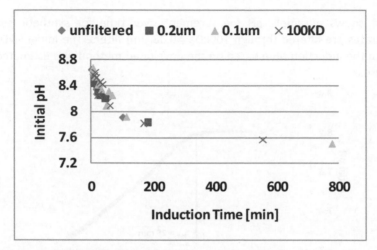

Figure 5: Effect of filter pore size on induction time of synthetic SWRO concentrate at a mixing speed of 150 rpm and temperature of 20 °C

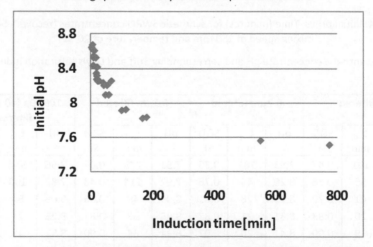

Figure 6: The relation between initial pH and induction time for all data points for synthetic SWRO concentrate at a mixing speed of 150 rpm and temperature of 20 °C

In Figure 6, a single logarithmic line appeared to fit all experimental data with a correlation factor of 0.95 and the produced relation suggests that immediate precipitation will not occur except when the solution pH is increased to ca. 9.08.

The relation between the induction time and the S&DSI (Figure 7) showed that the induction time was greater than 750 min at S&DSI = 0.12 (pH =7.51). This induction period is 100 times shorter (6-7min) at the highest supersaturation level applied of S&DSI $\approx$ 1.2 (pH = 8.67).

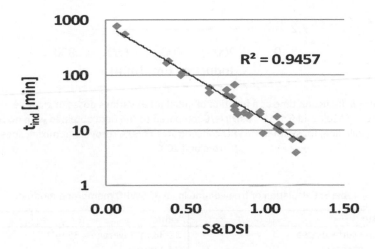

Figure 7: Induction time (minutes) vs. Stiff & Davis stability index for synthetic SWRO concentrates (recovery 50%) with all filtered and unfiltered experimental data (mixing speed of 150 rpm and temperature of 20 °C).

The effect of particle addition on the induction time

Immediate precipitation was observed when the solution was subjected to a suspension of high surface area glass particles (0.042 m^2/L) (Figure 8). Immediately, visual turbidity and a rapid drop in pH (from an initial pH of 8.15 down to 7.4) was observed after adding particles (1mL) and an induction time could not be practically measured (less than 1 minute). The pH at which the induction time was instantaneous was (8.15) and is much lower than the value suggested earlier (Figure 6) of 9.08. In the case of a particle concentration of $8*10^{12}$ /L (0.005 m^2/L), the induction time was 59 minutes and 78 minutes for an initial pH of 8 and 7.93, respectively. When compared to induction times measured for 'particle free' (filtered) solutions, these results are 30% lower than the expected values of 90 and 120 minutes. A visible observation of the glass reactor after the experiment showed that the reactors were scratched probably as a result of the glass bead addition in combination with vigorous stirring. Scratching of the reactor wall may have resulted in the creation of new (fresh) particles that may stimulate the nucleation process [15].

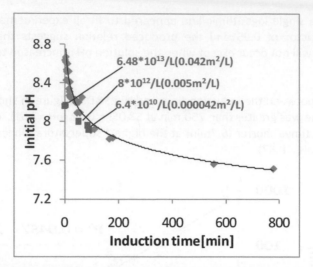

Figure 8: Induction time as a function of initial pH at various doses of glass beads of $6.48*10^{13}$; $8*10^{12}$ and $6.4*10^{10}$ particle/L compared to the data obtained with no particle addition (solid line) for synthetic SWRO concentrates of 50% recovery at mixing speed of 150 rpm and 20°C.

Table 5: Calculation of free volume in an 8" SWRO membrane module.

Parameter	Value
Membrane surface Area	32-38 m^2 (average = 35 m^2)
Thickness of spacer	0.52 mm
Membrane diameter	0.2 m
Length of the membrane module	1 m
Volume between the two membrane sheets	35m*0.52*10^{-3}m*1m=0.0182 m^3

In real SWRO plants, the surface area of the membrane in direct contact with water inside a module is nearly 1.9 m^2/L. The calculations are based on a membrane surface area of 32-38 m^2 and a spacer thickness of 0.52 mm for an 8 inch membrane with 1 m Length. The total free volume occupied by the feed water is nearly 18.2 L. The surface area of the RO membrane (1.9 m^2/L) is more than 3600 times greater than the surface area of particles in real seawater ($5.22*10^{-4}$ m^2/L) as suggested by Park et al, 2009 [4]. This may suggest that membrane scaling of SWRO modules is highly likely to be influenced by the membrane surface in direct contact with the concentrate, rather than particles present in the feed water.

The effect of mixing speed on the induction time

The mixing speed used in the experiments was varied between 10; 50 and 300 rpm and results of the induction time were compared with results employing a mixing speed of 150 rpm (the blue squares in the log relation plotted in Figure 9). The results shown in Figure 9 suggest that the mixing speed had almost no effect on the induction time in the tested range of mixing speed and saturation. These results agree with Tariner, 1981 [27] and Wiechers et al, 1975 [28] who concluded that the induction time was independent of the mixing rates applied. Others confirmed the effect of mixing on induction time [21, 32-33]. This effect was minimal at high supersaturation ratios (mainly in the homogenous range) [29]. This may suggest that the supersaturation values tested for the effect of mixing (S_a= 20-40) lies in the homogenous nucleation zone rather than the heterogeneous zone. However, further research is needed to prove this hypothesis.

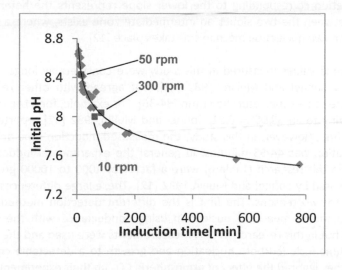

Figure 9: Induction time as a function of initial pH at mixing speeds of 300; 50 and 10rpm (red squares). Data obtained at 150rpm (blue squares fitted with black line) are also plotted for comparison

The nucleation mechanism

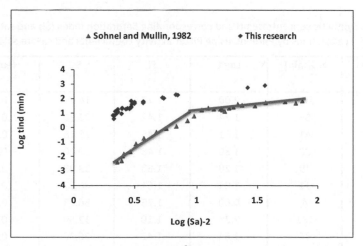

Figure 10: The relation between Log t_{ind} & Log^{-2} (Sa) for this research compared with Sohnel & Mullin, 1982 [12].

The experimental results of this study were compared to the classification developed by Sohnel & Mullin, 1982 [12] to identify the zones of homogenous and heterogeneous nucleation for $CaCO_3$. According to Sohnel & Mullin, 1982 [12], if Log (t_{ind}) is plotted against $Log^{-2} S_a$ for a wide range of saturations, two different slopes can be identified (Figure 10). The range of saturation corresponding to the steeper slope was assigned the homogenous zone of nucleation while the range of supersaturation corresponding to the lower slope represents the heterogeneous zone. In between the two slopes an intermediate zone exists, where a transition between the two nucleation mechanisms takes place [12].

Induction time values measured in this study were considerably longer than that reported by Sohnel and Mullin, 1982 [12], but agreed with other researchers' results at the same supersaturation ratio [34-36]. For example, for a Log $(S_a)^{-2}$ of 0.5 corresponding to an S&DSI = 0.81, Sohnel and Mullin,1982 [12] reported 0.1 sec induction time. However, in this study, the measured induction time at the same supersaturation was 44-55 minutes. In general the experimental induction times measured in this research (Table 6) were a factor of 1000 to 10000 greater than those measured by Sohnel and Mullin,1982 [12]. These large differences might be attributed to two reasons. The first is the different detection methods applied. Sohnel and Mullin measured nucleation using conductance with the stop flow technique, but in this research the pH measurements were used and the identified induction time included both nucleation and growth to a detectable crystal size. Secondly, they ignored the effect of atmospheric CO_2 on their experimental results. In chapter 2, much shorter induction time was found if the CO_2 interference is ignored.

The experimental data generated in this study were plotted in Figure 10 , and it can be seen that most of our data lie in the homogenous and the transition nucleation zones as defined by Sohnel and Mullin, 1982 [12].

Table 6: Synthetic concentrate pH and corresponding Saturation index (SI) and saturation ratio (S_a) calculations by PhreeqC using Pitzer activity coefficients and calcite solubility.

Initial pH	t_{ind}[min]	Log t	SI	S_a	$(Log S_a)^{-2}$
7.91	101	2.00	1.17	14.79	0.731
8.20	55	1.74	1.42	26.30	0.496
8.23	41	1.61	1.45	28.18	0.476
8.27	20	1.30	1.48	30.20	0.457
8.34	19	1.28	1.53	33.88	0.427
8.52	11	1.04	1.67	46.77	0.359
8.64	4	0.60	1.75	56.23	0.327
7.83	181	2.26	1.10	12.59	0.826
8.20	44	1.64	1.42	26.30	0.496
8.25	28	1.45	1.46	28.84	0.469
8.31	21	1.32	1.51	32.36	0.439

8.41	20	1.30	1.59	38.90	0.396
8.43	9	0.95	1.60	39.81	0.391
7.51	778	2.89	0.80	6.31	1.563
7.92	114	2.06	1.18	15.14	0.718
8.09	49	1.69	1.33	21.38	0.565
8.25	68	1.83	1.46	28.84	0.469
8.41	22	1.34	1.59	38.90	0.396
8.53	13	1.11	1.67	46.77	0.359
8.62	7	0.85	1.73	53.70	0.334
8.67	7	0.85	1.77	58.88	0.319
7.56	554	2.74	0.84	6.92	1.417
7.82	168	2.23	1.09	12.30	0.842
8.09	60	1.78	1.33	21.38	0.565
8.25	24	1.38	1.46	28.84	0.469
8.52	18	1.26	1.67	46.77	0.359
8.54	10	1.00	1.68	47.86	0.354
8.60	13	1.11	1.72	52.48	0.338

When ($Log^{-2} S_a$) was calculated for the mixing speed experiments (Figure 9), results showed values of 0.67, 0.61 and 0.39 for mixing speeds of 10, 300 and 50 rpm, respectively. These values appear to lie in the homogenous range as defined earlier by Sohnel and Mullin,1982 [12]. The minor effect of mixing on induction time may be explained by the fact that homogenous nucleation dominated, because it was reported that the effect of mixing is minor when homogenous nucleation is the governing mechanism. On the contrary, the effect on the Induction time is expected to be greater when heterogeneous nucleation predominates [21, 27-28, 32-33].

3.6. Conclusions

- Supersaturated SWRO concentrate (recovery of 50%) was produced by dissolving pure salts (99.9%) in ultra-pure water to avoid the presence of foreign particles in the supersaturated solution. In addition the supersaturated solutions were pre-filtered with either 0.2 µm; 0.1 µm and 100 kDa (0.03 µm) filters to remove any particles present in the solution. No effect of pre-filtration with 0.2µm; 0.1µm and 100 kDa (0.03 µm) was observed on the induction time for the synthetic SWRO case with initial S&DSI of 0.12 to 1.28. A possible explanation for this result may be that particles smaller than 0.03 µm affect induction time. However more research is required to verify this hypothesis.

- The induction time for $CaCO_3$ in synthetic seawater concentrate was not affected by the range of mixing rates tested in this study (10-300 rpm). For example, at 50 rpm the induction time was 14 minutes compared to 15 minutes in the case of 150 rpm. According to the definition of how to differentiate between homogenous and heterogeneous nucleation mechanisms proposed by Sohnel and Mullin, 1982 [12], most of the experimental data produced in this study lies in the homogenous and intermediate zones. This may explain the minor effect of mixing on the induction time.

3.7. List of symbols

A	Function of Pre-exponential factor ($s^{-1}m^{-3}$)
$A*$	Deby Huckel constant ($L^{2/3}mol^{-1/2}$)
a_c	Activity
A_f	Free area for precipitation at a given particle size (m^2)
Alk	Alkalinity of solution (mole/L)
B	Constant expressed in equation 1.7 ($L^{3/2}mol^{-1/2}nm^{-1}$)
$f(\theta)$	Factor differentiating heterogeneous and homogenous nucleation
I	The ionic strength (mol/L)
IAP	Ionic activity product (mol^2/L^2)
J	Nucleation rate (nuclei/min/cm^3)
Js	Steady state nucleation rate (nuclei/min/cm^3)
k	Boltzmann constant (J/K)
K_{sp}	Solubility product ($mole^2/L^2$)
K_{a2}	Second acidity constant (mole/L)
pH_c	Concentrate pH
pH_s	Equilibrium pH
SI	Supersaturation Index
S_a	Supersaturation ratio
T	Absolute temperature in Kelvin
t_{ind} (min)	Induction time in minutes unless mentioned otherwise
v	Number of ions into which a molecule dissociates
V_m	Molecular volume (cm^3/mole)

Greek letters

β	Geometric factor
γ_+	Cation activity coefficient

γ_- Anion activity coefficient

γ^s Surface energy (J/m^2)

Ω Pre-exponential factor in the nucleation rate equation ($s^{-1}m^{-3}$)

3.8. References

1.   Potts, D.E., R.C. Ahlert, and S.S. Wang, A critical review of fouling of reverse osmosis membranes. Desalination, 1981. 36: p. 235–264.

2.   Yiantsios, S.G. and A.J. Karabelas, An assessment of silt density index based on RO membrane colloidal fouling experiments with iron oxide particles. Desalination, 2002. 151: p. 229-238.

3.   Hoof, S.C.J.M.V., J.G. Minnery, and B. Mack, Performing a membrane autopsy. Desalination Water Reuse, 2002. 11: p. 40-46.

4.   Park, K., et al., Measurement of size and number of suspended and dissolved nonoparticles in water for evaluation of colloidal fouling in RO membranes. Desalination, 2009. 238: p. 78–89.

5.   Kubota, N. and J.W. Mullin, A kinetic model for crystal growth from aqueous solution in the presence of impurity. Journal of Crystal Growth, 1995. 152: p. 203-208.

6.   Abdel-Aal, E., M. Rashad, and H. El-Shall, Crystallization of calcium sulfate dihydrate at different supersaturation ratios and different free sulfate concentrations. Cryst. Res. Technol., 2004. 39: p. 313–321.

7.   Zuddas, P. and A. Mucci, Kinetics of calcite precipitation from seawater: II the influence of the ionic strength. Geochimica et Cosmochimica Acta, 1998. 62: p. 757-766.

8.   Chen, T., A. Neville, and M. Yuan, Assessing the effect of Mg^{2+} on $CaCO_3$ scale formation–bulk precipitation and surface deposition. Journal of Crystal Growth, 2005. 275: p. 1341-1347.

9.   Gledhill, D.K. and J.W. Morse, Calcite dissolution kinetics in Na–Ca–Mg–Cl brines. Geochimica et Cosmochimica Acta, 2006. 70: p. 5802-5813.

10.   Gledhill, D.K. and J.W. Morse, Calcite solubility in Na–Ca–Mg–Cl brines. Chemical Geology, 2006. 233: p. 249-256.

11.   Kirkova, E. and M. Djarova, On the Kinetics of Crystallization of Zinc Oxalate Dihydrate by Precipitation Industrial Crystallization, 1979. 78: p. 81-89.

12.   Sohnel, O. and J.W. Mullin, Precipitation of calcium carbonate. Journal of Crystal Growth, 1982. 60: p. 239-250.

13.   Bansal, B., H. Muller-Steinhagen, and X.D. Chen, Effect of suspended particles on crystallization fouling in plate heat exchangers. ASME, 1997. 119: p. 568-574.

14.   Thompson, C., Investigating the fundamentals of drug crystal growth using atomic force microscope. 2003, The University of Nottingham.

15.   Snoeyink, V. and D. Jenkins, Water Chemistry. 1980, New York: John Wiley & Sons.

16.   ASTM, Calculation and Adjustment of the Stiff and Davis Stability Index for Reverse Osmosis. 2001, ASTM International: West Conshohocken, PA, United States.

17.   Elfil, H. and H. Roques, Role of hydrate phases of calcium carbonate on the scaling phenomenon. Desalination, 2001. 137: p. 177-186.

18.   El-Manharawy, S. and A. Hafez, Water type and guidelines for RO system design. Desalination, 2001. 139(1-3): p. 97-113.

19.   Schippers, J., Membrane filtration- Reverse Osmosis, Nanofiltration & Electrodialysis. 2003, Delft: UNESCO-IHE.

20.   Sheikholeslami, R., Assessment of the scaling potential for sparingly soluble salts in RO and NF units. Desalination, 2004. 167: p. 247-256.

21.   Sohnel, O. and J. Garside, Precipitation basis principals and industrial applications. 1992, Oxford: Butterworth-Heinemann.

22.   Stiff, H.A. and L.E. Davis, A method for predicting the tendency of oil field waters to deposit calcium carbonate. Petroleum transactions, 1952. 195: p. 213-216.

23.   Waly, T., et al., Will calcium carbonate really scale in seawater reverse osmosis? Desalination and water treatment, 2009. 5: p. 252-256

24.   Boerlage, S., Scaling and particulate fouling in membrane filtration system, in Sanitary Engineering. 2002, IHE: Delft.

25.   Mullin, J.W. and S. Zacek, The precipitation of potassium aluminium sulphate from aqueous solution Journal of Crystal Growth, 1981. 53: p. 515-518.

26.   Yu-zhu, S., et al., Seeded Induction Period and Secondary Nucleation of Lithium Carbonate. The Chinese Journal of Process Engineering, 2009. 9(4): p. 652-660.

27.   Trainer, D., A study of calcium carbonate crystal growth in the present of a calcium complexing agent. 1981, The Institute of Paper Chemistry: Wisconsin.

28.   Wiechers, H.N.S., P. Sturrock, and G.V.R. Morais, Calcium carbonate crystallization kinetics. Water Research, 1975. 9: p. 835-845.

29.   Sohnel, O. and J.W. Mullin, Influence of mixing on batch precipitation Crystal Research and Technology 1987. 22(10): p. 1235 - 1240.

30.   Sohnel, O., Electrolyte crystal-aqueous solution interfacial tensions from crystallization data. Journal of Crystal Growth, 1982. 57: p. 101-108.

31.   Walton, A.G., In dispersion of powders in liquids. 2nd ed. 1973, London: Appl. Sci. Publ.

32.   Kuboi, R., et al. Mixing Effects in Double-Jet and Single-Jet Precipitation. in World Congress III of Chemical Engineering. 1986. Tokyo.

33.   Mohanty, R., et al., Charactarizing the product crystals from a mixing tee process. AIChE journal, 1988. 34(12): p. 2063-2068.

34.   Butt, F.H., F. Rahman, and U. Baduruthamal, Hollow fine fiber vs. spiral-wound reverse osmosis desalination membranes Part 2: Membrane autopsy. Desalination 1997. 109: p. 83-94.

35.   Hannachi, A., et al. A new index for scaling assessment. in IDA World Congress-Maspalomas. 2007. Gran Canaria.

36.   Koutsoukoas, P. and C. Kontoyannis, Precipitation of calcium carbonate in aqueous solutions. J. Chem. Soc., Faraday Trans.I, 1984. 80: p. 1181-1192.

Chapter 4

On the induction time of CaCO$_3$ in high ionic strength synthetic seawater

Based on parts of:

Waly, T.; Kennedy, M.D.; Witkamp, G.J.; Amy, G. and Schippers, J.C. On the induction time of CaCO$_3$ in high ionic strength synthetic seawater. Desalination and Water Treatment, accepted.

Waly, T.; Saleh, S.; Kennedy, M.D.; Witkamp, G.J.; Amy, G. and Schippers, J.C. Will calcium carbonate really scale in seawater reverse osmosis? Desalination and Water Treatment, 2009. 5 p252-256

Waly, T.; Saleh, S.; Kennedy, M.D.; Witkamp, G.J.; Amy, G. and Schippers, J.C. Will calcium carbonate really scale in seawater reverse osmosis? In proceeding of the EDS (Ed.) EuroMed conference, 2008. Dead Sea, Jordan

4.1.   Abstract

In practice, the $CaCO_3$ scaling potential in SWRO plants can be determined by identifying the critical phase of precipitation and the mechanism of nucleation involved. This research aimed to investigate the induction times of $CaCO_3$ as a function of the saturation and ionic strength for synthetic seawater with a focus on the concentrate bulk solution. The investigation verified theoretical calculations by using experiments to determine the critical phase of $CaCO_3$ and the mechanism involved. The experimental procedure utilized in this research is a sensitive and stable pH meter with accuracy of 0.001 pH units. Induction time experiments were performed with synthetic concentrates have different ionic strengths of 0.054, 1.12, 1.34, 1.61 mole/L. The synthetic solutions had the same Ca^{2+} and HCO_3^- concentrations as SWRO concentrate, based on seawater from the Gulf of Oman, with a recovery of 30% (I = 0.054, 1.12 mole/L) and 50% (I = 1.34, 1.61 mole/L). In these sets of experiments, induction time was defined as the time needed for the solution pH to drop 0.03 pH units. Results showed a strong correlation between the logarithm of the induction time and the saturation and the Stiff & Davis saturation indices. For low ionic strength solution (I=0.054 mole/L), the S&DSI matched the SI when the solubility of calcite was incorporated into the SI calculations. On the contrary, for high ionic strength water (I = 1.12, 1.34, 1.61 mole/L) the S&DSI matched the SI using the solubility of vaterite in its calculation. These findings indicate that vaterite and not calcite maybe the precipitating phase in seawater. This expectation was confirmed with microscopic analysis of the formed crystals at the end of the induction time experiments (24hrs) as vaterite was found in the solution instead of calcite.

The study of the mechanism of nucleation showed that in the range of saturations used, three different trends were recognized describing three different nucleation mechanisms; namely homogenous, heterogeneous and a one characterized by an intermediate surface tension. The nucleation mechanism involved appear to be related to the initial supersaturation of the solution. Homogenous nucleation predominates when the initial supersaturation is greater than that of the hexahydrated $CaCO_3$ phase. On the other hand, heterogeneous nucleation is the dominant mechanism when the solution is supersaturated with regard to vaterite. In between is a transition zone where the solution is supersaturated with respect to monohydrated $CaCO_3$ but undersaturated with respect to hexahydrated $CaCO_3$. The calculations of the apparent surface energy showed values ranging from 15-94 mJ/m^2 which is consistent with literature data for vaterite using the free drift method. Results suggest that in SWRO systems with pH values of around 8.0, the nucleation mechanism will most probably be heterogeneous, which encourages further studies to determine the effect of the membrane surface on the nucleation mechanism.

Keywords: nucleation, membrane, desalination, surface energy, phase

4.2.   Background

Introduction

In seawater reverse osmosis (SWRO) plants where 25-50% of the seawater is converted into fresh water, scaling problems are considered a constant potential in plant design and operation. There are many types of scale that may form in a plant, but the most common scalants in SWRO are $CaCO_3$ and $Mg(OH)_2$ (second pass) [1]. Pre-treatment using acid is widely used for $CaCO_3$ scaling prevention in SWRO plants. For the purpose of optimizing acids doses, solution stability after its saturation limits are exceeded (induction time) [2] need to be studied. The aim of this research is to determine the effect of salinity on the induction time of $CaCO_3$ in SWRO concentrates.

Seawater salts mainly comprise 7 main ions: Ca^{2+}, Mg^{2+}, Na^+, K^+, Cl^-, HCO_3^- and SO_4^{2-} as shown in Table 1. The pH of the seawater usually ranges between 7.8 (e.g. the Red Sea) to 8.3 (e.g. the Atlantic Ocean). Lower pH values (< 7.8) may be found if beach wells are the main feed water source.

Table 1: Seawater composition (intake of desalination plant in the Gulf of Oman)

Parameter	Average	unit
Alkalinity total as $CaCO_3$	120	mg/L
Ammonium	0.03	mgN/L
Calcium	474	mg/L
conductivity	60	mS/cm
Carbonate	23	mg/L
Sodium	12,244	mg/L
Hardness non carbonate($CaCO_3$)	6,718	mg/L
Hardness total ($CaCO_3$)	6,838	mg/L
Magnesium	1,356	mg/L
pH	8.1	
Phosphate	0.1	mgP/L
Silicate	0.13	mg/L
Chloride	21,535	mg/L
Sulphate	2,772	mg/L
Density	1.03	g/cm3
TDS	39,017	mg/L
Ionic strength	0.76	mole/L

Solubility of salts

When an ionic compound is added to water, it will usually dissolve in the solution as its ions and complexes. If the activity of an ionic compound added to a volume is in excess, an equilibrium is reached when the number of ions entering the solution from the solid compound is equal to the number of ions leaving the solution to the solid compound according to the following reaction represented in equation 1 [3]:

$$CaCO_3 \rightleftharpoons Ca^{2+} + CO_3^{2-}$$ 1.

The first requirement for precipitation or scale formation is supersaturation of the solution with respect to the scaling salt. If the solubility is exceeded based on the maximum amount of salt soluble in a solution (at a given temperature) precipitation may occur [2, 4-7]. The equilibrium thermodynamic solubility product (K_{sp}) varies with temperature wherein alkaline scale solubility decreases as the temperature increases [8]. It is also noted that the rate of scaling may be enhanced by surface roughness, hydraulic conditions, as well as the surface charge [9].

The thermodynamic solubility product can be measured in a laboratory using the seeded growth technique [4, 7] in addition, activity coefficients and common ion effect are incorporated as shown in equations 2 and 3 [4, 10-12].

$$K_{sp} = [Ca^{2+}][CO_3^{2-}]\gamma_+\gamma_-$$ 2.

$$pK_{sp} = pK_{so} + log\,\gamma_+ + log\,\gamma_-$$ 3.

The solubility and acidity constants of $CaCO_3$ in particular are very sensitive to ionic strength and ion pairing affecting the saturation calculations for $CaCO_3$ in seawater. Pitzer et al [13-14] have taken into account the effect of ion pairing on the calculations of the activity coefficients for high ionic strength waters up to, e.g. 6 moles/L. Using this approach in solubility product calculations helps the researcher to model real scalant behaviour in saline waters [4].

Common methods used to determine the precipitation potential of $CaCO_3$:

In seawater, $CaCO_3$ saturation is calculated by the use of different indices such as the Stiff & Davis Stability Index (S&DSI) [15]; the Saturation Index (SI) and the Saturation Ratio (S_a). These indices were developed to predict the possibility to precipitate of sparingly soluble salts, in general, except S&DSI which is developed specifically for $CaCO_3$.

Stiff & Davis Stability Index (S&DSI)

Developed in 1952, the S&DSI [16] is widely used for calcium carbonate scaling potential in higher ionic strength solutions similar to that of sea water (equations 4 & 5) [15].

$$S\&DSI = pH - pH_s$$ 4.

$$pH_s = pCa + pHCO_3^- + k \qquad\qquad 5.$$

Where pHs is the saturation pH , pCa is the $-Log_{10}$ Ca, $pHCO_3^-$ is the $-Log_{10}$ HCO_3^- and k is a factor compensating for salinity and temperature.

The S&DSI was developed after experimental work to take into account the ionic strength in saturation calculations. The ionic strength limit in the S&DSI index ranges between 0 – 3.6 mole/L [16] and is limited to temperature ranges from 0 °C to 50°C, making it a very useful tool for calcium carbonate scaling determination in seawater reverse osmosis systems. A positive S&DSI [16] indicates the tendency to form calcium carbonate scale, while a negative value indicates no tendency for scaling formation. One of the main problems of this index is that it can not be extrapolated out of its salinity and temperature ranges.

Stiff & Davis carried out their experiments for 24 hours (time dependent) at different levels of ionic strength and at three temperatures, namely 0, 30, 50 °C. The measuring technique of Stiff and Davis [16] is based on increasing the salinity by adding NaCl up to 20% (200,000 mg/L). They measured the limit at which precipitation starts for each different ionic strength value and for each of the three temperatures they used.

The advantage of this technique is that it incorporates the change dissociation constants due to the increase in temperature and salinity. Although not known, the solubility product of the precipitating phase is incorporated in k value of the S&DSI. Some disadvantages of using S&DSI that it ignores the effect of seawater salts other than NaCl e.g. Mg^{2+}, SO_4^{2-}, K^+, etc. which may affect the accuracy of the predicted saturation using S&DSI. Furthermore, it ignores the effect of natural organic matter present in all natural waters. Another point is that there is some arbitrariness of obtaining the solubility in their measurements as it is measured after 24hrs from the experimental start.

Saturation Index (SI)

This saturation index predicts the scaling potential of sparingly soluble salts taking into consideration the interaction between ions. The index incorporates the activity coefficients in its activity calculations which takes into account the effect of ionic complexation due to high salinity (equations 6-9). For SWRO concentrates it is preferable to use Pitzer model rather than Davis or extended Debye Huckel models in calculating the activity coefficients as the former takes into account the ionic specific interaction [2, 4, 14, 17-20].

$$SI = Log \left(\frac{[Ca^{2+}]\gamma_{ca^{2+}}[CO_3^{2-}]\gamma_{co_3^{2-}}}{K_{sp}} \right) \qquad\qquad 6.$$

$$pH = -Log\,[H]\gamma_{H^+} = pK_{a2} + log\,\frac{[CO_3^{2-}]\gamma_{CO_3^{2-}}}{[HCO_3^-]\gamma_{HCO_3^-}} \qquad\qquad 7.$$

$$SI = Log\left(\frac{[Ca^{2+}]\gamma_{Ca^{2+}}.K_{a2}[HCO_3^{2-}]\gamma_{HCO_3^-}}{K_{sp}[H]\gamma_{H^+}}\right) \qquad\qquad 8.$$

$$SI = pH - (p[Ca^{2+}] + p[HCO_3^-] + pK_{a2} - pK_{sp} + p\gamma_{Ca^{2+}} \\ + p\gamma_{HCO_3^-}) \qquad\qquad 9.$$

Where $\gamma_{Ca^{2+}}, \gamma_{CO_3^{2-}}, \gamma_{HCO_3^-}$ and γ_{H^+} are the activity coefficients of calcium, carbonate, bicarbonate and hydrogen, respectively, log X is $log_{10}X$, K_{a2} is the second acidity constant and K_{sp} is the thermodynamic solubility product.

When compared to S&DSI (equations 10 and 11)

$$S\&DSI = pH - (p[Ca^{2+}] + p[HCO_3^-] + k) \qquad\qquad 10.$$

$$k = -pK_{sp} + pK_{a2}^T + p\gamma_{Ca^{2+}} + p\gamma_{HCO_3^-} \qquad\qquad 11.$$

Values of S&DSI greater than zero mean that it is more likely that salt will scale while values less than zero suggests that the solution is undersaturated with respect to this specific salt [17].

Saturation Ratio (S$_a$)

Although S_a is less widely used if compared to the SI, it adopts the same concept for saturation calculations as shown in equations 12 and 13 [2].

$$S_a = \frac{[Ca^{2+}]\gamma_{Ca^{2+}}[CO_3^{2-}]\gamma_{CO_3^{2-}}}{K_{sp}} \qquad\qquad 12.$$

Values of S_a greater than one mean that it is more likely that the salt will scale while values less than one mean that the solution is undersaturated with respect to this specific salt [2].

$$SI = Log\,S_a \qquad\qquad 13.$$

Mechanism of nucleation

The mechanism of nucleation is divided into homogenous or heterogeneous nucleation [21]. Homogeneous nucleation is where the formation of the solid phase is not influenced by the presence of any solid phase, and heterogeneous nucleation is where the formation of new solid phase particles is catalyzed by the presence of a foreign solid phase [2]. The formed nucleus grows further to the crystalline phase while The formation of stable crystals needs an aging step for the formation of the final product [22-23].

Homogenous nucleation

For the formation of a solid phase cluster comprising N molecules and having surface A_N, the change of Gibbs energy accompanying this change can be defined as showed in equation 14 [2]

$$\Delta G_{hom} = -N\emptyset + A_N \gamma_s = \frac{k_v r^3}{\vartheta}\emptyset + k_a r^2 \gamma_s \qquad\qquad 14.$$

Where ΔG_{hom} is the change in Gibbs energy in the homogenous zone, N is the number of molecules, A_N is the nuclei surface area γ_s represents the surface energy, $\emptyset$ is the affinity (equation 20), K_v volume geometric factor (equals 1 for cubic shapes), K_a area geometric factor (equal 1 for cubic shapes), ϑ molecular volume and r is the nuclei radius

Evaluating the critical radius in the equation of the Gibbs free energy gives the maximum free Gibbs energy (also called energy barrier), in which the probability of its growth and its decay are equal (equation 15) [2].

$$\frac{d\Delta G_{hom}}{dN} = -\emptyset + \frac{A_N}{dN}\gamma_s = 0 \qquad\qquad 15.$$

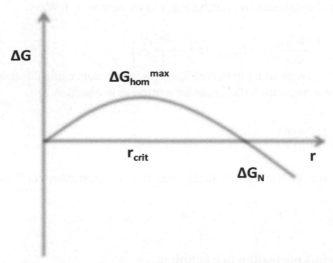

Figure 1: Free energy of nucleation as a function of cluster size [24]

And the critical radius r_{crit} in such cases is found as shown in equation 16 and represented in Figure 1

$$r_{crit} = \frac{2 \vartheta k_a \gamma_s}{3 k_v \emptyset}$$ 16.

By substitution and rearrangement

$$\Delta G_{hom}^{max} = \frac{\beta \vartheta^2 \gamma_s^3}{\emptyset^2}$$ 17.

Thus, using the geometrical constant K_a and K_v, the geometric expression β can be expressed as shown in equation 18:

$$\beta = \frac{4 k_a^3}{27 k_v^2}$$ 18.

And consequently the number of molecules forming the critical nuclei can be written as in equation 19.

$$N = \frac{V}{\vartheta} = \frac{k_v r^3}{\vartheta} = \frac{2 \beta \vartheta^2 \gamma_s^3}{\emptyset^3}$$ 19.

Where V is the nuclei volume which can be related to the supersaturation (S_a)as in equation 20

$$\emptyset = v k_b T Ln S_a$$ 20.

And the rate of nucleation (J) in such cases can be written as follows:

$$J = \Omega_{hom} exp \left(\frac{-\Delta G_{hom}}{k_b T} \right) = \Omega_{hom} exp \left(\frac{-\beta \vartheta^2 \gamma_s^3}{k_b T \emptyset^2} \right)$$ 21.

Where T is the temperature in degree Kelvin, k_b is the Boltzmann constant, and the pre-exponential expression (Ω_{hom}) can be written as in equation 22

$$\Omega_{hom} = \frac{D}{d^5 N} \left(\frac{\Delta G_{hom}}{3 \pi k_b T} \right)^{0.5}$$ 22.

Where d is the inter-planer distance and D is the diffusion coefficient in the solution.

Heterogeneous nucleation in a solution

The catalytic effect of a solid phase presence on nucleation is explained by the resulting decrease in the energy barrier to nucleation if the nucleus forms on the surface of the solid phase [2]. ΔG_{het} can be described in terms of homogeneous parameters, together with a single additional parameter $f(\theta)$ (equation 23-25), which represents the contact angle between the crystalline deposit and the solid substrate [25].

$$\Delta G_{het} = f(\theta)\Delta G_{hom} \qquad\qquad 23.$$

Where

$$f(\theta) = \frac{(1 - \cos\theta)(2 + \cos\theta)}{4} \qquad\qquad 24.$$

Where θ represents the contact angle between the solid surface and the spherical nucleus [25]

The rate of heterogeneous nucleation can be evaluated from the change in Gibbs energy in a similar way to that used for homogeneous nucleation.

$$J = \Omega_{het}exp\left(\frac{-\Delta G_{het}}{k_b T}\right) = \Omega_{het}exp\left(\frac{-\beta\vartheta^2\gamma_s^3 f(\theta)}{k_b T \emptyset^2}\right) \qquad\qquad 25.$$

Where Ω_{het} is the pre-exponential expression for heterogeneous nucleation and $\Omega_{het} < \Omega_{hom}$.

In practice, the period of metastability preceding crystallization process is commonly indicated as the induction time (t_{ind}) [2, 4, 26-27]. The induction time calculation is inversely related to the nucleation rate as in equations 26 and 27 [2].

$$t_{ind} \propto J^{-1} \qquad\qquad 26.$$

Where the nucleation steady state rate can be expressed as

$$t_{ind} = \frac{1}{\Omega}exp\left(\frac{\beta\vartheta^2\gamma_s^3 f(\theta)}{k_b T \emptyset^2}\right) \qquad\qquad 27.$$

Which can be written in a simplified way as follows:

$$logt_{ind} = \frac{B}{(T)^3.log^2 S_a} - A \qquad\qquad 28.$$

Where B and A are constants and can be expressed as in equations 29 and 30

$$B = \left[\frac{\beta\vartheta^2\gamma_s^3 f(\theta)}{v^2(2.3k_b)^3}\right] \qquad\qquad 29.$$

And

$$A = log\Omega$$ 30.

Experimentally the measured induction time may also include growth to a detectable size [2].The induction time in such case is proportional to both nucleation (J) and growth rates (G) as shown in equation 31 [28-30].

$$t_{ind} \propto (G^3 . J)^{\frac{-1}{4}}$$ 31.

For a limited range of supersaturation, no change in the nucleation mechanism is expected and the relation can be written as a linear fit between the logarithm of the induction time and that of the saturation ratio [2, 31-33].

$$logt_{ind} = k - n. log(S_a)$$ 32.

Where k is a constant with no physical meaning and n is the order.

CaCO$_3$ phases

According to equation 32 the induction time depends strongly on the saturation (S_a). The seawater saturation is known to be affected by the ionic strength, affecting the solubility of $CaCO_3$ and acidity constants. High ionic strength increases the CO_2 solubility as well as the acidic constants leading to, if not taken into consideration, an overestimation of the saturation [11, 34-41]. Some researchers have claimed the formation of other forms of $CaCO_3$ other than calcite [2, 18, 34-35, 42-45].

According to Jansen [46] and El Manharawy and Hafez [47] that seawater is supersaturated with regard to calcite (the most stable form of $CaCO_3$); this saturation is halved after the first 100-200 meters of the ocean surface [46-48]. It is claimed that calcite is formed in the upper ocean layer whereas it is consumed in the lower layer by means of $CaCO_3$ -consuming micro organisms [46].

According to El Manharawy and Hafez [47] and after examining 15,000 natural water samples from different natural water sources indicated that surface oceans and seas are much less supersaturated than was previously thought. Their study depended on the measurement of the total alkalinity of the natural water and compared it with the maximum total alkalinity found at the same chloride content. The results suggest that seawaters are undersaturated with regard to calcite.

These results can be seen alongside the results of Morales et al, 1996 [49] where they proved that the higher the ratio of Ca^{2+} to water HCO_3^-, the longer the induction time. His results suggested a factor of 4 as the optimum ratio for obtaining the longest induction time which is (by coincidence) the same ratio that is found naturally in seawater.

Jansen, H., 2001 [46] came to the same conclusion in his study pointing out that seawater is undersaturated if compared to the solubility of aragonite and calcite. This undersaturation increases by the increase with depth and this is mainly due to the presence of $CaCO_3$-consuming organisms in the deep sea.

The formation of hydrated $CaCO_3$ forms

Ostwald's rule of stage (1897) states that an unstable system does not necessarily transform directly into the most stable state, but may be preceded by one which most closely resembles its own, i.e. into another transient stage which possesses greater energy than that of the stable phase [2, 50].

Based on Ostwald's rule of stage, the activation energy for the formation of $CaCO_3$ must exceed that of the least stable form. In this specific case and depending on temperature, the first precipitating phase will be either amorphous calcium carbonate (> 30 °C) or ikaite ($CaCO_3.6H_2O$) (<25 °C) [45] (Figure 2) or amorphous and monohydrated depending on the mechanism of nucleation [42, 51] as amorphous $CaCO_3$ (100 times higher in solubility than calcite) formation is associated with homogenous nucleation and monohydrated $CaCO_3$ (40 times higher than calcite) with heterogeneous nucleation.

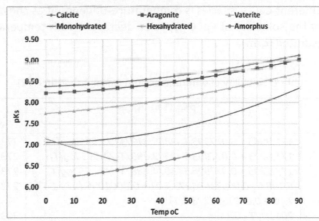

Figure 2: The relation between the solubility of $CaCO_3$ phases and temperature[42] .

By the formation of the first precipitating phase, the reaction continues until the formation of calcite which is the most stable form of $CaCO_3$ (Figure 3). This was explained by the kinetics of transformation, which means that the phase which has the highest formation rate is more likely to form (e.g. monohydrated $CaCO_3$ to vaterite) than that which has more favourable thermodynamics (e.g. monohydrated to calcite) [2, 52].

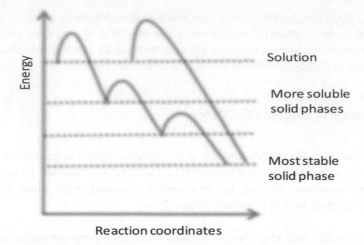

Figure 3: The activation energy needed vs. the crystalline phase, where lower barrier is the most soluble phase while the higher barrier is the least soluble, the stepwise model on the left exerts the least energy and is more likely to take place than the right one [53].

Calcite as the most stable form of $CaCO_3$ has been extensively studied by different researchers [2, 11, 54]. Other compounds such as aragonite, vaterite, amorphous, monohydrated and hexahydrated calcium carbonate have been studied less. The solubility of different $CaCO_3$ species is temperature-dependent as shown in Table 2 [18]. Normally the solubility of $CaCO_3$ species decreases with temperature except for the hexahydrated form.

Table 2: Solubility of different calcium carbonate forms [42]

Species	pK$_{sp}$ (T in °K)	Range of validity
Calcite	$171.9065 + 0.077993.T - 2839.319/T - 71.595.log(T)$	273.1<T<263.1°K
Aragonite	$171.9773 + 0.077993.T - 2903.293/T - 71.595.log(T)$	273.1<T<263.1°K
Vaterite	$172.1295 + 0.077996.T - 3074.688/T - 71.595.log(T)$	273.1<T<263.1°K
Amorphous	$6.1987 + 0.0053369.(273.1-T) + 0.0001096.(273.1-T)^2$	273.1<T<228.1°K
Monohydrated	$7.05 + 0.000159.(273.1-T)$	273.1<T<223.1°K
Hexahydrated	$2011.1/T - 0.1598$	283.1<T<298.1°K

Amorphous calcium carbonate (ACC) is the most unstable form, and is often formed in spherical shapes with a diameter less than a micrometer. The ACC initially formed is transformed within a few minutes to a mixture of more stable phases of calcium carbonate. The transformed carbonates are vaterites and calcites at low temperatures (14-30°C) and aragonites and calcites at higher temperatures (60-80°C). At intermediate temperatures (40-50°C) the formation of all three species was observed [42].

Monohydrate calcium carbonate (MCC) has a hexagonal shape crystal. The synthesis of this mineral requires the presence of magnesium as well as organic material. Both in nature and in the laboratory, this mineral readily crystallizes at a temperature near $0°C$, yet rapidly decomposes to anhydrous forms at higher temperatures (T > $10°C$) [18].

The $CaCO_3$-CO_2-H_2O system was studied by Elfil et al, (2004) [43] with different $CaCO_3$ species. They focused on the different hydrated forms of $CaCO_3$. The results showed that the metastable zone lies between the K_{sp} values of amorphous $CaCO_3$ and that of the monohydrated $CaCO_3$.

Up until now major seawater membrane manufacturers have underestimated the scaling limit of $CaCO_3$ in SWRO plants. This overestimation of scaling potential is shown in their recommendations for acid and antiscalant addition to the feed water. Their calculations for $CaCO_3$ saturation are based on the S&DSI and Langelier saturation index (LSI). Several researchers and field studies have suggested that seawater may be undersaturated and not oversaturated at all, as previously believed. Others have claimed that Mg^{2+}; SO_4^{2-} and humic substances may act as natural antiscalants [2, 34, 39-41, 55-56].

The effect of pressure on the saturation with $CaCO_3$ was studied intensively by petroleum researchers [57-58]. One of these studies by Dyer and Graham, 2002 [57] showed that there is a decrease in supersaturation due to pressure increase. This may be due to the increase in CO_2 solubility with increasing pressure [35].

4.3.   Materials

pH meter

The induction time measurements using pH were performed by a highly sensitive pH meter (Eutech pH 6000) with an accuracy of 0.001 pH units. pH can be measured online using the manufacturer's (Eutech) software or offline by using the instrument memory for sample storage. The measuring interval can be adjusted to as low as every 30 seconds.

Reactors

The pH probe was fitted in the top of the air-tight double-jacketed glass reactor with a volume of 3 litres (Applikon) (Figure 4), and equipped with a double-paddled shaft mechanical stirrer. The mixing rate can be varied from 0 to 1200 rpm using an electronic controller (Applikon) linked to the motor of the mixer. The reactor can be filled either manually or mechanically using a diaphragm pump with an average filling speed of 4 L/min.

After each experiment, cleaning was employed with 0.2 molar HNO_3 for 30 minutes with a flow of 0.15 L/min to dissolve any formed crystals. The reactor was then flushed with de-mineralized water for 15 minutes with a flow rate of 3 L/min before the next experiment.

Figure 4: The reactors assembly and setup

Synthetic seawater concentrate preparation

The preparation of ultra-pure water starts by delivering water as the raw water source where it passes through a series of treatment steps to decrease the organic and inorganic particle content in the feed water. The product water had a conductivity and total organic carbon (TOC) of 0.8 μS/cm and 3 μg/L, respectively. The TOC was measured using a TOC analyzer with a detection limit of 0.5 μg/L.

The synthetic seawater concentrate used in the experiments are prepared in stages. Firstly, $NaHCO_3$ solution was prepared by dissolving $NaHCO_3$ salt (Table 3) in ultrapure water. Secondly, a $CaCl_2$. solution was prepared by dissolving $CaCl_2.2H_2O$. Finally, NaCl salt was dissolved in the previously prepared $CaCl_2.2H_2O$ solution to adjust the salinity values of the prepared synthetic concentrate to the desired levels.

To ensure a complete dissolution of the reagents, the preparation step involves dissolving the solutions on a 1 L batch basis. The salt was added to the ultra-pure water in a measuring flask. The flask was then closed and shaken manually for 2 minutes after which 2 hours of solution mixing took place on a magnetic stirrer. Mixing was performed at an average speed of 400 rpm and at a room temperature of 20 °C.

Table 3: Salt reagents used in the experimental synthetic seawater concentrate preparation

Reagent	Form	♀Supplier	Purity
NaCl	Salt	J.T.Baker	99.5-99.9 %
$CaCl_2.2H_2O$	Salt	MERCK	99.9
$NaHCO_3$	Salt	MERCK	99.9

The induction time experiments were initiated by adding the $NaHCO_3$ solution into the reactor followed by the addition of 0.2 M NaOH solution for pH correction (if needed). The mixing rate during addition is kept at 150 rpm to prevent the formation of local areas with higher saturation than desired one. Finally the $CaCl_2$. + NaCl solution was added with a rate of 0.2 L/min while maintaining a mixing speed of 150 rpm to ensure proper mixing and to prevent the formation of local saturation zones. The addition was performed through fine nozzles located 3 cm from the reactor's base accompanied by vigorous mixing to ensure proper distribution of the solution when added. The two reacting solutions were added on a 1:1 volume basis.

4.4.   Methods

Induction time measurements

The pH change was monitored over a period of 1000 minutes, and the induction time was defined as a pH decrease of at least 0.03 pH units. This decrease is equivalent to less than 0.1 - 0.27 mg/L of precipitated $CaCO_3$ (30-50 times lower than the total precipitated) depending on the added HCO_3^- concentration which is constant for each ionic strength tested but fixed for each recovery. The mixing rate was kept constant at 150 rpm using an electronic controller linked to the mixing motor and the start of the experiment was defined as the time of complete addition of the reacting solutions.

Solutions concentrations

Table 4: The experimental solution's ionic strength and the corresponding calcium and bi-carbonate content

Recovery	%	30%		50%	
Ionic strength	Mole/L	0.054	1.12	1.34	1.61
Ca^{2+}	mg/L	677	677	948	948
HCO_3^-	mg/L	209	209	293	293

Four experimental matrixes were used in the induction time experiments with ionic strengths of I = 0.054, 1.12, 1.34, 1.61 mole/L (Table 4). For the lower ionic strength range the Ca^{2+} and the HCO_3^- content were 667 and 209 mg/L, respectively. In the higher range these amounts were increased to 948 and 293 mg/L, respectively. These amounts of Ca^{2+} and HCO_3^- are equivalent to those found in SWRO concentrates using Gulf of Oman water at recovery rates of 30% and 50%.

Calculation of SI using PhreeqC

The Phreeqc program was used to calculate the activity product by incorporating the activity coefficient to account for the ionic complexation due to the increase in salinity. The SI calculations in this study used the ionic activity product from Phreeqc and equations presented in Table 2 for the calculation of different $CaCO_3$ phase saturations.

4.5.   Results

Table 5 represents the results for the experimental data between the initial pH, initial SI and induction time at four different ionic strength values of I = 0.054, 1.12, 1.34, and 1.61 mole/L (Table 5).

Table 5: The induction time experimental results at different ionic strength for synthetic seawater concentrates

I mole/L	Initial pH	t_{ind} (min)	Log t_{ind}	Initial S&DSI	log IAP	T °K	Initial SI (calcite)
	7.84	173	2.24	1.20	-7.17	298	1.31
	7.83	192	2.28	1.19	-7.18	298	1.30
	8.12	41	1.61	1.48	-6.91	298	1.57
	8.26	37	1.57	1.62	-6.8	298	1.68
	8.28	23	1.36	1.64	-6.79	298	1.69
0.054	8.4	22	1.34	1.76	-6.69	298	1.79
	8.41	17	1.23	1.77	-6.68	298	1.80
	8.56	15	1.18	1.92	-6.57	298	1.91
	8.58	11	1.04	1.94	-6.56	298	1.92
	8.81	11	1.04	2.17	-6.45	298	2.03
	8.75	5	0.70	2.11	-6.41	298	2.07
	7.82	678	2.83	0.03	-7.64	293	0.81
	8.04	374	2.57	0.25	-7.43	293	1.02
	8.28	184	2.26	0.49	-7.22	293	1.23
	8.32	240	2.38	0.53	-7.19	293	1.26
	8.42	114	2.06	0.63	-7.1	293	1.35
1.12	8.50	138	2.14	0.71	-7.0	293	1.41
	8.63	100	2.00	0.84	-6.94	293	1.51
	8.68	73	1.86	0.89	-6.91	293	1.54
	8.79	41	1.61	1.00	-6.84	293	1.61
	8.88	25	1.40	1.09	-6.78	293	1.67

	8.98	26	1.41	1.19	-6.72	293	1.73
	9.04	40	1.60	1.25	-6.69	293	1.76
	9.10	22	1.34	1.31	-6.66	293	1.79
	7.93	114	2.06	0.54	-7.19	298	1.29
	8.23	60	1.78	0.84	-6.93	298	1.55
	8.33	49	1.69	0.94	-6.85	298	1.63
1.34	8.42	48	1.68	1.03	-6.79	298	1.69
	8.52	21	1.32	1.13	-6.71	298	1.77
	8.62	22	1.34	1.23	-6.65	298	1.83
	8.79	17	1.23	1.40	-6.55	298	1.93
	7.51	778	2.89	0.07	-7.58	293	0.87
	7.56	554	2.74	0.12	-7.54	293	0.91
	7.82	168	2.23	0.38	-7.29	293	1.16
	7.83	181	2.26	0.39	-7.28	293	1.17
	7.91	101	2.00	0.47	-7.21	293	1.24
	7.92	114	2.06	0.48	-7.20	293	1.25
	8.00	85	1.93	0.56	-7.16	293	1.29
	8.04	62	1.79	0.60	-7.10	293	1.35
	8.09	49	1.69	0.65	-7.05	293	1.40
	8.09	60	1.78	0.65	-7.05	293	1.40
	8.20	55	1.74	0.76	-6.96	293	1.49
	8.20	44	1.64	0.76	-6.96	293	1.49
	8.23	41	1.61	0.79	-6.93	293	1.52
	8.25	28	1.45	0.81	-6.92	293	1.53
	8.25	24	1.38	0.81	-6.92	293	1.53
1.61	8.27	20	1.30	0.83	-6.90	293	1.55
	8.31	21	1.32	0.87	-6.87	293	1.58
	8.34	19	1.28	0.90	-6.85	293	1.60
	8.41	20	1.30	0.97	-6.79	293	1.66
	8.41	22	1.34	0.97	-6.79	293	1.66
	8.43	9	0.95	0.99	-6.78	293	1.67
	8.43	14	1.15	0.99	-6.77	293	1.68
	8.52	11	1.04	1.08	-6.71	293	1.74
	8.52	18	1.26	1.08	-6.71	293	1.74
	8.53	13	1.11	1.09	-6.71	293	1.74
	8.54	10	1.00	1.10	-6.70	293	1.75
	8.60	13	1.11	1.16	-6.66	293	1.79
	8.62	7	0.85	1.18	-6.65	293	1.80
	8.64	4	0.60	1.20	-6.63	293	1.82
	8.67	7	0.85	1.23	-6.61	293	1.84

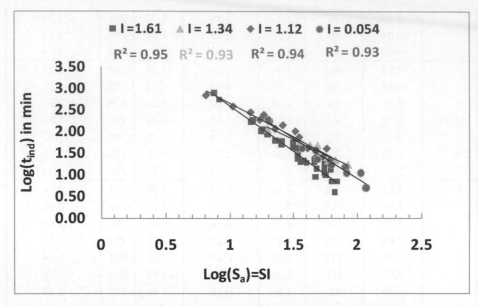

Figure 5: Log t_{ind} (minutes) vs. Log (Sa) for high Ionic strength water

The relation between Log t_{ind} and Log (S_a) represented in Figure 5 showed a correlation factor ranging between 0.93 - 0.95 when Equation 32 describing the nucleation mechanism for a limited range of saturation was used.

Adopting the same concept, the relation between S&DSI and the logarithm of the induction time (Figure 6) was plotted. The results showed a different pattern than that found in Figure 5 between SI and the logarithm of the induction time. In Figure 6 two different zones can be identified where the first contains the low ionic strength induction time experiments (I = 0.054 mole/L) while the second contains the higher ionic strength experiments (I = 1.12, 1.34, 1.61 mole/L). In the first zone (I = 0.054), the S&DSI values were nearly 0.5-1.1 units of magnitude less if compared to the S&DSI values obtained at the same induction time values at high ionic strength solutions (I = 1.12, 1.34, 1.61 and represented in Table 5).

Results in Table 5, Figures 5 and 6 shows that the value of the S&DSI and SI at low ionic strength solutions are comparable. On the contrary, at higher ionic strength solutions values the S&DSI values were nearly 0.5 units of magnitude less compared to that of SI. A possible explanation is that the S&DSI and SI are using similar solubility product value (calcite in such case) but different solubility product value at high ionic strength but this hypothesis needs to be further investigated.

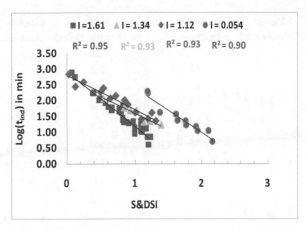

Figure 6: Comparison relationship between Log t$_{ind}$ (minutes) vs. S&DSI for I = 0.054, 1.12, 1.34, 1.61 mole/L at 25°C except for I = 1.61 mole/L at 20°C

The Solubility product used by the S&DSI

A relation between SI, S&DSI and pH was established for different CaCO$_3$ phases to determine the K$_{sp}$ value incorporated in the S&DSI (Figure 7-10). In Figures 8, 9, 10 for I = 1.61, 1.34, 1.12 mole/L, the results suggest that to have the SI and the S&DSI value are comparable if the solubility of vaterite is used in the SI calculation. On the other hand, at the lower ionic strength of I = 0.054 the SI and the S&DSI values were nearly equal when the K$_{sp}$ was in the range of calcite. It is worth mentioning that the change in the phase solubility used by Stiff & Davis, 1952 [16] may be attributed to their experimental procedure which depend on shaking the supersaturated solution and measuring the solubility after 24 hrs at different ranges of ionic strength.

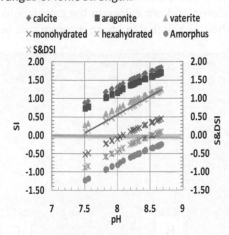

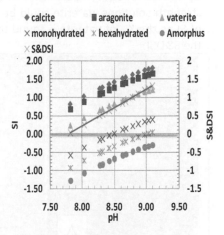

Figure 7: Comparison relationship between SI and S&DSI (crosses with blue line) vs. pH for different phases at I = 1.61

Figure 9: Comparison relationship between SI and S&DSI (crosses with blue line) vs. pH for different phases at I = 1.12

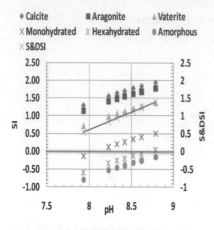

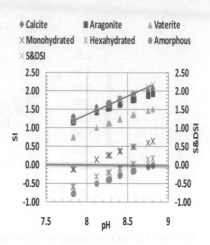

Figure 8: Comparison relationship between SI and S&DSI (crosses with blue line) vs. pH for different phases at I = 1.34

Figure 10: Comparison relationship between SI and S&DSI (crosses with blue line) vs. pH for different phases at I = 0.054

Samples were taken after 24 hrs from the reactor at the end of the induction experiments for solutions with I = 0.054 and analyzed in wet conditions under an electron microscope (Figure 11). Results showed a mixture of vaterite (hexagonal shape) and calcite (rhombus) crystals. These findings were very similar to those found by Clarkson et al, 1997 [59] where a mixture of vaterite and calcite was found as an end product of their nucleation experiments. The same was reported when vaterite is formed at normal seawater temperatures and transformed into calcite due to its instability [53]. The presence of vaterite at the end of the experiment suggests that SI calculated using the K_{sp} of calcite and S&DSI calculations for low ionic strength solutions are overestimating the supersaturation by a factor of 0.5 compared to that calculated using calcite solubility. At high ionic strength solutions, vaterite could be considered and not calcite in the calculations of SI as the K_{sp} of vaterite is found to be the most probable phase incorporated in the S&DSI.

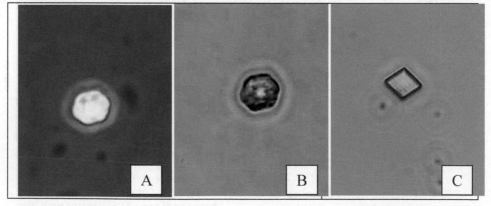

Figure 11 water (I = 0.054) and the solution saturation S_a of 42 at 25°C, **A &B**- the vaterite hexagonal crystal **C**- calcite crystal shape

Homogenous and heterogeneous nucleation

Although Equation 27 suggests a straight line correlation between Log t_{ind} and $T^{-3}Log^{-2}(S_a)$, Figure 12 showed that a low correlation factor (0.76-0.85) was found for synthetic solutions with I = 1.12, 1.34, 1.61 mole/L. On the contrary, the correlation factor was 0.95 for experiments of I = 0.054. This may be attributed to the change in the nucleation trend (nucleation mechanism involved) for the higher ionic strength solutions but not in the low ionic strength solution.

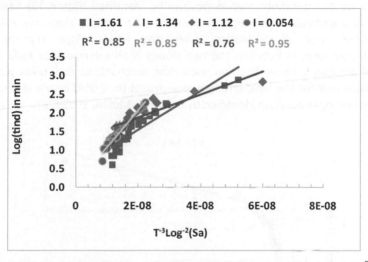

Figure 12: Log t_{ind} (minutes) vs. SI for low and high ionic strength water at 25 °C.

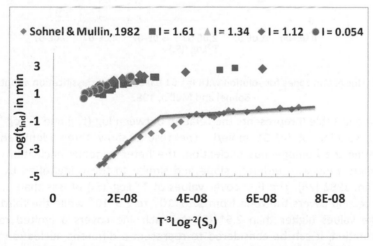

Figure 13: Induction time results of this research compared to Sohnel and Mullin, 1982 data for the classification of homogenous and heterogeneous mechanism

The previous hypothesis was confirmed by comparing our research data against the data of Sohnel and Mullin 1982 [26] (Figure 13); the same pattern was found. Figure 13 shows that the induction time measured in this research is 1000-100 time longer compared to that of Sohnel & Mullin, 1982 [26]. This may be attributed to the differences in identifying the induction time, where in Sohnel & Mullin, 1982 [26] research induction time represents only nucleation but in our research, induction time included both nucleation and growth to detectable size.

According to Sohnel & Mullin, 1982 [26], if Log (t_{ind}) is plotted against $T^{-3}Log^{-2} S_a$ for a wide range of saturations, two slopes can be identified (Figure 13) where each represents a particular nucleation mechanism. The steeper slope represents the homogenous zone of nucleation while the lower slope represents the heterogeneous one. In between the two slopes is an intermediate zone where a smooth transition between the two nucleation mechanisms may takes place. The results show that for low ionic strength experiment (I= 0.054) nearly all the results lies in the homogenous zone identified by Sohnel & Mullin, 1982 [26].

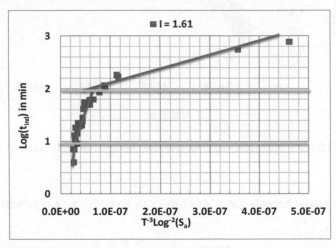

Figure 14: Nucleation zones for solution with I=1.61 based on the classification adopted from Sohnel and Mullin, 1982

Figure 14 and Table 6 represents the relation between log (t_{ind}) and $T^{-3}Log^{-2} S_a$ in a synthetic solution of I=1.61 mole/L. The results show three identified zones representing the homogenous nucleation, the heterogeneous nucleation and an intermediate zone with different slope and similar to those identified by Sohnel and Mullin, 1982 [26]. The first cover values of $T^{-3}.Log^{-2}(S_a)$ of less than $1.25*10^{-8}$ and the second covers the zone from $1.25*10^{-8}$ to $2.0*10^{-8}$ while the third covers the range values higher than $2.6*10^{-8}$. As each line covers a limited range of supersaturation, it can be considered to represent a different nucleation surface energy (different $\gamma_s^3 f(\theta)$) [2]. The steeper line represents the homogenous nucleation zone while the line with the lower slope represents the heterogeneous one.

Figure 15 shows the initial SI using different K_{sp} values for hydrated and unhydrated $CaCO_3$ phases as a function of log (t_{ind}). Figure 16 and Table 6 was divided into three zones based on the initial saturation compared to the $CaCO_3$ phases. At the first zone (Log t_{ind} <1.1) the experimental solution was initially supersaturated with respect to the unhydrated phases (calcite, aragonite, vaterite) but not to the hydrated phases (amorphous, hexahydrated, monohydrated). In the second zone (1.79 <Log t_{ind} > 1.1), the solution was supersaturated with regard to monohydrated $CaCO_3$ but undersaturated with respect to hexahydrated and amorphous $CaCO_3$. In the third zone (Log t_{ind} > 1.79), the solution was supersaturated with hexahydrated $CaCO_3$ but not with amorphous $CaCO_3$. In this identified third zone, the increase in the initial pH of the solution in order to increase SI hardly increases the solution saturation and therefore, amorphous $CaCO_3$ saturation could not be reached in our experiments.

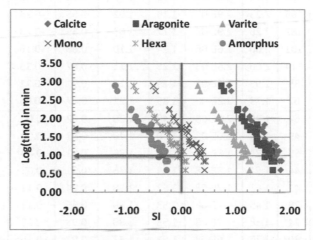

Figure 15: Relation between the log (t_{ind}) and the SI using the solubility of different phases of $CaCO_3$ for solution with I=1.61 mole/L

The borders of the three identified zones are identical to the points of slope change in Figure 14 and Table 6 and hence the mechanism of nucleation. The results suggest that homogeneous nucleation predominates only when the initial solution saturation exceeds that of hexahydrated $CaCO_3$. The intermediate zone, exists when the initial saturation in the range between monohydrated $CaCO_3$ and hexahydrated $CaCO_3$. In the lower saturation range (lower than monohydrated $CaCO_3$) Heterogeneous nucleation predominates. Therefore these results may suggest that the mechanism of nucleation is related not only to the solution saturation but must also be combined with the phase of the $CaCO_3$.

Table 6: Nucleation mechanism classification based on I=1.61 mole/L

	pH	t_{ind} (min)	Log t_{ind}	$T^{-3}.Log^{-2}$ (S_a)	K_{sp}					
					8.45	8.31	7.87	7.05	6.70	6.35
					Calcite	Aragonite	Vaterite	Mono-hydrated	Hexa-hydrated	Amorphous
					SI					
Group 1 Heterogeneous	7.51	778	2.89	5.2E-08	0.87	0.73	0.29	-0.53	-0.88	-1.23
	7.56	554	2.74	4.8E-08	0.91	0.77	0.33	-0.49	-0.84	-1.19
	7.82	168	2.23	2.9E-08	1.16	1.02	0.58	-0.24	-0.59	-0.94
	7.83	181	2.26	2.9E-08	1.17	1.03	0.59	-0.23	-0.58	-0.93
	7.91	101	2.00	2.6E-08	1.24	1.10	0.66	-0.16	-0.51	-0.86
	7.92	114	2.06	2.5E-08	1.25	1.11	0.67	-0.15	-0.50	-0.85
	8.00	85	1.93	2.4E-08	1.29	1.15	0.71	-0.11	-0.46	-0.81
	8.04	62	1.79	2.2E-08	1.35	1.21	0.77	-0.05	-0.40	-0.75
Group 2 intermediate	8.09	60	1.78	2.0E-08	1.40	1.26	0.82	0.00	-0.35	-0.70
	8.09	49	1.69	2.0E-08	1.40	1.26	0.82	0.00	-0.35	-0.70
	8.20	55	1.74	1.8E-08	1.49	1.35	0.91	0.09	-0.26	-0.61
	8.20	44	1.64	1.8E-08	1.49	1.35	0.91	0.09	-0.26	-0.61
	8.23	41	1.61	1.7E-08	1.52	1.38	0.94	0.12	-0.23	-0.58
	8.25	28	1.45	1.7E-08	1.53	1.39	0.95	0.13	-0.22	-0.57
	8.25	24	1.38	1.7E-08	1.53	1.39	0.95	0.13	-0.22	-0.57
	8.27	20	1.30	1.6E-08	1.55	1.41	0.97	0.15	-0.20	-0.55
	8.31	21	1.32	1.6E-08	1.58	1.44	1.00	0.18	-0.17	-0.52
	8.34	19	1.28	1.5E-08	1.60	1.46	1.02	0.20	-0.15	-0.50
	8.41	20	1.30	1.4E-08	1.66	1.52	1.08	0.26	-0.09	-0.44
	8.41	22	1.34	1.4E-08	1.66	1.52	1.08	0.26	-0.09	-0.44
	8.43	14	1.15	1.41E-08	1.68	1.54	1.10	0.28	-0.07	-0.42
	8.43	9	0.95	1.4E-08	1.67	1.53	1.09	0.27	-0.08	-0.43
	8.52	11	1.04	1.3E-08	1.74	1.60	1.16	0.34	-0.01	-0.36
	8.52	18	1.26	1.3E-08	1.74	1.60	1.16	0.34	-0.01	-0.36
	8.53	13	1.11	1.3E-08	1.74	1.60	1.16	0.34	-0.01	-0.36
Group 3 Homogenous	8.54	10	1.00	1.3E-08	1.75	1.61	1.17	0.35	0.00	-0.35
	8.60	13	1.11	1.2E-08	1.79	1.65	1.21	0.39	0.04	-0.31
	8.62	7	0.85	1.2E-08	1.80	1.66	1.22	0.40	0.05	-0.30
	8.64	4	0.60	1.2E-08	1.82	1.68	1.24	0.42	0.07	-0.28
	8.67	7	0.85	1.2E-08	1.84	1.70	1.26	0.44	0.09	-0.26

Accordingly, following the same procedure, Figures 16 &17 were drawn for a synthetic solution with I=1.12 using the experimental results represented in Table 7. The concept appears to be valid in the heterogeneous zone where the solution is supersaturated with respect to the unhydrated phases but undersaturated with the hydrated ones. When the monohydrated saturation was surpassed, a different slope was found representing the same transitional period as the one found earlier in Figures 14 and 15. The final stage of homogenous nucleation was not identified in this figure due to the lack of data (there were only two points).

Table 7: The nucleation mechanism classification based on I=1.12 mole/L

	pH	t_{ind} (min)	Log t_{ind}	$T^{-3}.Log^{-2}(S_a)$	K_{sp}					
					8.45	8.31	7.87	7.05	6.70	6.35
					Calcite	Aragonite	Vaterite	Mono-hydrated	Hexa-hydrated	Amorphous
Group 1	7.82	678	2.83	6.0E-08	0.81	0.67	0.23	-0.59	-0.94	-1.29
	8.04	374	2.57	3.8E-08	1.02	0.88	0.44	-0.38	-0.73	-1.08
	8.28	184	2.26	2.6E-08	1.23	1.09	0.65	-0.17	-0.52	-0.87
	8.32	240	2.38	2.5E-08	1.26	1.12	0.68	-0.14	-0.49	-0.84
	8.42	114	2.06	2.2E-08	1.35	1.21	0.77	-0.05	-0.40	-0.75
Group 2	8.50	138	2.14	2.0E-08	1.41	1.27	0.83	0.01	-0.34	-0.69
	8.63	100	2.00	1.7E-08	1.51	1.37	0.93	0.11	-0.24	-0.59
	8.68	73	1.86	1.7E-08	1.54	1.40	0.96	0.14	-0.21	-0.56
	8.79	41	1.61	1.5E-08	1.61	1.47	1.03	0.21	-0.14	-0.49
	8.88	25	1.40	1.4E-08	1.67	1.53	1.09	0.27	-0.08	-0.43
	8.98	26	1.41	1.3E-08	1.73	1.59	1.15	0.33	-0.02	-0.37
Group 3	9.04	40	1.60	1.3E-08	1.76	1.62	1.18	0.36	0.01	0.34
	9.10	22	1.34	1.2E-08	1.70	1.65	1.21	0.39	0.04	-0.31

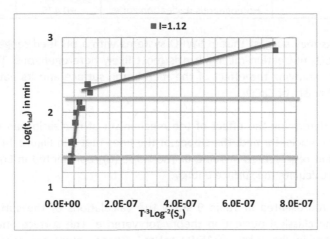

Figure 16: Nucleation zones for solution with I=1.12 based on the classification adopted after Sohnel and Mullin, 1982

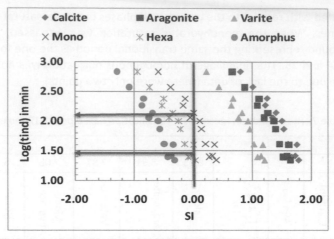

Figure 17: Relation between the log (t_{ind}) and the SI using the solubility of different phases of CaCO$_3$ for solution with I=1.12 mole/L

Table 8: The intersection and slope values (A &B) in Equations 28, 29 and 30 for the three identified nucleation mechanism zones for solution of I =1.61.

Ionic strength	I = 1.61	A	B
I = 1.61	Heterogeneous nucleation zone	-1.7346	3.00E+06
	Intermediate nucleation zone	-0.3806	2.00E+07
	Homogenous nucleation zone	+1.1132	8.00E+08
I = 1.34	Heterogeneous nucleation zone	--	--
	Intermediate nucleation zone	-0.5829	3.00E+07
	Homogenous nucleation zone	--	--
I = 1.12	Heterogeneous nucleation zone	-1.9654	1.00E+07
	Intermediate nucleation zone	-0.2259	9.00E+07
	Homogenous nucleation zone	--	--
I = 0.054	Heterogeneous nucleation zone	--	--
	Intermediate nucleation zone	-0.6011	3.00E+07
	Homogenous nucleation zone	+0.475	8.00E+07

These results may suggest that in SWRO systems with a pH feed range of around 8.0, the CaCO$_3$ nucleation mechanism is most likely heterogeneous. This suggest the need for studying the effect of the membrane surface and its nature on the scaling mechanism of CaCO$_3$.

In order to determine the effect of the mechanism of nucleation on the surface energy of the nuclei formed, the slopes of the lines found in Figure 14 & 16 were calculated and represented in Table 8. The slope (B) represented in Equation (29) was used to calculate the surface energy.

The results represented in Table 9 show the calculations of the surface energy based on the classical nucleation theory for vaterite. The surface energy values obtained were in the range of 15-94 mJ/m^{-2}. These values are consistent with literature values for vaterite when the heterogeneous nucleation mechanism predominates [51, 54, 60-68] except for the higher value of 94 mJ/m^2 which is consistent with theoretical literature values of (90-141 mJ/m^2) for surface energy of vaterite when homogenous nucleation mechanism predominates. It is worth mentioning that the uncertainty in the shape factor (β) of the blocks building the formed nuclei (either spheres or cubes) will affect the resultant surface energy value e.g. using the shape factor for spheres instead of cubes will result in a surface energy value of 114 mJ/m^2 instead of 94 mJ/m^{-2}. Furthermore, the value of surface energy was reported to be affected by the experimental technique used where the free drift method (similar to what is used in this work) normally results in a much lower surface energy value compared to the constant composition method [62, 66-67].

Table 9: The calculations of the apparent surface energy using Equation 29.

I mole/L	B	k_a	k_v	β	ρ	m	υ	ϑ	$\gamma_s^3 f(\theta)$	γ_s(apparent) J/m^2
	3.00E+06	1	6	32	2660	0.1	2	6.24E-29	3.08E-06	0.015
1.61	2.00E+07	1	6	32	2660	0.1	2	6.24E-29	2.05E-05	0.027
	8.00E+08	1	6	32	2660	0.1	2	6.24E-29	8.21E-04	0.094
	0.0E+00	1	6	32	2660	0.1	2	6.24E-29	0.00E+00	0.000
1.34	3.00E+07	1	6	32	2660	0.1	2	6.24E-29	3.08E-05	0.031
	0.0E+00	1	6	32	2660	0.1	2	6.24E-29	0.00E+00	0.000
	1.00E+07	1	6	32	2660	0.1	2	6.24E-29	1.03E-05	0.022
1.12	9.00E+07	1	6	32	2660	0.1	2	6.24E-29	9.23E-05	0.045
	0.00E+00	1	6	32	2660	0.1	2	6.24E-29	0.00E+00	0.000
	0.00E+00	1	6	32	2660	0.1	2	6.24E-29	0.00E+00	0.000
0.054	3.00E+07	1	6	32	2660	0.1	2	6.24E-29	3.08E-05	0.031
	8.00E+07	1	6	32	2660	0.1	2	6.24E-29	8.21E-05	0.043

4.6.   Conclusions

1. The S&DSI showed similar values when compared with SI the solubility of calcite in its calculations for low ionic strength synthetic solution (I=0.054 mole/L). On the contrary, for high ionic strength water (I = 1.12, 1.34, 1.61 mole/L) the S&DSI was similar to SI using the solubility of vaterite in its calculation. The results suggested that vaterite is most likely the precipitating phase and not calcite. Incorporating the solubility of vaterite and not calcite in the SI calculation will result in a decrease in SI by about 0.5 units compared to that using the solubility of calcite.

2. The relation between log t_{ind} and $T^{-3} Log^{-2}$ (S_a) can be used to identify the nucleation mechanism involved for $CaCO_3$. Three different slopes were found where the steepest slope represents the homogenous nucleation and the shallower line representing the heterogeneous nucleation zone. An intermediate zone with different line slope and consequently a different surface energy was recognized.

3. The nucleation mechanism involved in the induction time experiment is closely related to the initial synthetic solution saturation. A homogenous precipitation appears to predominate when the initial solution saturation exceeded that of hexahydrated $CaCO_3$ (ikaite) and the heterogeneous nucleation appears to predominate when the initial saturation is less than the monohydrated but higher than vaterite. In between, an intermediate zone where the initial saturation is below the ikaite but higher than the hexahydrated.

4. The surface energy calculations using the classical nucleation theory result in surface energy values of 15-94 mJ/m^2 which is consistent with values in literature for spontaneous precipitation of $CaCO_3$. The uncertainty due to the shape factor value will result in a 20% increase in the calculated surface energy if spherical building blocks are used in the calculations compared to these values calculated using cubic building block shapes.

4.7.   List of symbols

A	Function of Pre-exponential factor (s^{-1}m^{-3})
A*	Deby Huckel constant (L$^{2/3}$mole$^{-1/2}$)
A_N	Nuclei surface area (m^2)
a_c	Activity
A_f	Free area for precipitation at given particle size (m^2)
Alk	Alkalinity of solution (mole/L)
B	Constant expressed (L$^{3/2}$mole^{-1}/m^{-1})
D	Diffusion coefficient in the solution
d	Interplaner distance in solid phase (m)
ΔG_{hom}	The change in Gibbs energy in the homogenous zone (J)
ΔG_{het}	The change in Gibbs energy in the heterogeneous zone (J)
G	Growth rate (mole/min/cm^3)
I	The ionic strength (mole/L)
IAP	Ionic activity product (mole^2/L^2)
J	Nucleation rate (nuclei/min/cm^3)
K_a	Area geometric factor
K_{a2}	Temperature corrected second acidity constant (mole/L)
k_b	Boltzmann constant (J/K)
K_{so}	Solubility at standard conditions (mole^2/L^2)
K_{sp}	Solubility product (mole^2/L^2)
K	Graphical obtained constant in S&DSI calculations
k_a	Area geometric factor
k_v	Volume geometric factor
Log	Log$_{10}$
N	Number of molecules

n	the order of nucleation
P	Pressure (psi)
pAlk	Negative the logarithm of alkalinity activity divided by its dimensions
pCa	Negative the logarithm of calcium activity divided by its dimensions
pH	Concentrate pH
pH_s	Equilibrium pH
S_a	Supersaturation ratio
SI	Supersaturation Index
r	Nuclei radius (m)
r^*	Critical nuclei radius (m)
T	Absolute temperature (Kelvin)
t_{ind}	Induction time in minutes unless mentioned otherwise (min)
V	The nuclei volume
υ	Number of ions into which a molecule dissociates
X	A constant with no physical meaning

Greek letters

$f(\theta)$	Factor differentiating heterogeneous and homogenous nucleation
β	Geometric factor
ϑ	Molecular volume (cm^3/mole)
$\gamma+$	Cation activity coefficient
$\gamma-$	Anion activity coefficient
γ_s	Surface energy (J/m^2)
Ω	Pre-exponential factor in the nucleation rate equation ($s^{-1}m^{-3}$)
ϕ	Pre-exponential factor in the nucleation rate equation

4.8.   References

1.   Schippers, J., Membrane filtration- Reverse Osmosis, Nanofiltration & Electrodialysis. 2003, Delft: UNESCO-IHE.
2.   Sohnel, O. and J. Garside, Precipitation basis principals and industrial applications. 1992, Oxford: Butterworth-Heinemann.
3.   Schippers, J., Desalination methods. 2003, Delft: UNESCO-IHE.
4.   Boerlage, S., Scaling and particulate fouling in membrane filtration system, in Sanitary Engineering. 2002, IHE: Delft.
5.   Jiang, S., Crystallization Kinetics in Polymorphic Organic Compounds. 2009, TUDelft: Delft.
6.   Van Der Weijden, R., Interactions between Cadmium and Calcite. 1995, Utrecht University: Utrecht.
7.   Van Der Leeden, M., The Role of Polyelectrolytes in Barium Sulphate Precipitation. 1991, TUDelft: Delft.
8.   Patel, S. and M. Finan, New antifoulants for deposite control in MSF and MED plants. Desalination, 1999. 124: p. 63-74.
9.   Lee, S., J.S. Choi, and Z.H. Lee, Behaviors of dissolved organic matter in membrane desalination. Desalination, 2009. 238: p. 109-116.
10.   Butler, J., Carbon Dioxide equilibria and their applications. second ed. 1982, California: Addison-Wesley.
11.   Sheikholeslami, R., Assessment of the scaling potential for sparingly soluble salts in RO and NF units. Desalination, 2004. 167: p. 247-256.
12.   Sohnel, O. and J. Garside, On Supersaturation Evaluation For Solution Growth. Journal of Crystal Growth, 1981. 54: p. 358-360.
13.   Pitzer, K.S., Thermodynamics of electrolytes. I. Theoretical basis and general equations. J. Phys. Chem., 1973. 77: p. 268-277.
14.   Pitzer, K.S. and G. Mayorga, Thermodynamics of electrolytes (III) Activity and osmotic coefficients for 2-2 electrolytes. J. of Soln. Chem., 1973. 3: p. 539-546.
15.   ASTM, Calculation and Adjustment of the Stiff and Davis Stability Index for Reverse Osmosis. 2001, ASTM International: West Conshohocken, PA, United States.
16.   Stiff, H.A. and L.E. Davis, A method for predicting the tendency of oil field waters to deposit calcium carbonate. Petroleum transactions, 1952. 195: p. 213-216.
17.   Sheikholeslami, R., Mixed salts--scaling limits and propensity. Desalination, 2003. 154(2): p. 117-127.
18.   Elfil, H. and H. Roques, Kinetics of the precipitation of calcium sulfate dihydrate in a desalination unit. Desalination 2003. 157: p. 9-16.
19.   Bromley, L.A., Thermodynamic properties of strong electrolytes. AIChE Journal, 1973. 19(2): p. 313-320.

20.   Wang, Y., Composite fouling of calcium sulphate and calcium carbonate in a dynamic seawater reverse osmosis unit. 2005, University of New South Wales: Sydney.

21.   Darton, E., Membrane chemical research: centuries apart. Desalination, 2000. 132: p. 121-131.

22.   Snoeyink, V. and D. Jenkins, Water Chemistry. 1980, New York: John Wiley & Sons.

23.   Sorber, A. and R. Valenzuela, Evaluation of an electrolytic water conditioning device for the elimination of water-formed scale deposits in domestic water systems, in center for research in water resources. 1982, Texas university: Austin.

24.   Voinescu, A.E., Biomimetic Formation of $CaCO_3$ Particles Showing Single and Hierarchical Structures, in Chemical and Pharmaceutical. 2008, University of Regensburg Regensburg

25.   Liu, Y. and G.H. Nancollas, Fluorapatite growth kinetics and the influence of solution composition. Journal of Crystal Growth, 1996. 165(1-2): p. 116-123.

26.   Sohnel, O. and J.W. Mullin, Precipitation of calcium carbonate. Journal of Crystal Growth, 1982. 60: p. 239-250.

27.   Sohnel, O. and J.W. Mullin, Influence of mixing on batch precipitation Crystal Research and Technology 1987. 22(10): p. 1235 - 1240.

28.   Avrami, M., Kinetics of phase change: I general theory. Journal of chemical physics, 1939. 7(12): p. 1103-1112.

29.   Avrami, M., Kinetics of phase change: II transformation time relations for random distribution of nuclei. Journal of chemical physics, 1940. 8(2): p. 212-224.

30.   Avrami, M., Kinetics of phase change: III granulation, phase change, and microstructure. Journal of chemical physics, 1941. 9(2): p. 177-184.

31.   Golubev, S.V., O.S. Pokrovsky, and V.S. Savenko, Unseeded precipitation of calcium and magnesium phosphates from modified seawater solutions. Journal of Crystal Growth, 1999. 205: p. 354-360.

32.   Sergei, V., O.S. Pokrovsky, and V.S. Savenko, Unseeded precipitation of calcium and magnesium phosphates from modified seawater solutions. Journal of Crystal Growth, 1999. 205: p. 354-360.

33.   Golubev, S.V., O.S. Pokrovsky, and V.S. Savenko, Homogeneous precipitation of magnesium phosphates from seawater solutions. Journal of Crystal Growth, 2001. 223: p. 550-556.

34.   Abdel-Aal, E., M. Rashad, and H. El-Shall, Crystallization of calcium sulfate dihydrate at different supersaturation ratios and different free sulfate concentrations. Cryst. Res. Technol., 2004. 39: p. 313–321.

35.   Al-Anezi, K., et al., Parameters affecting the solubility of carbon dioxide in seawater at the conditions encountered in MSF desalination plants. Desalination, 2008. 222: p. 548-571.

36.   Gledhill, D.K. and J.W. Morse, Calcite dissolution kinetics in Na–Ca–Mg–Cl brines. Geochimica et Cosmochimica Acta, 2006. 70: p. 5802-5813.

37.   Gledhill, D.K. and J.W. Morse, Calcite solubility in Na–Ca–Mg–Cl brines. Chemical Geology, 2006. 233: p. 249-256.

38.   Goyet, C. and A. Poisson, New determination of carbonic acid dissociation constants in seawater as a function of temperature and salinity. Deep sea research, 1989. 36(11): p. 1635-1654.

39.   Zuddas, P. and A. Mucci, Kinetics of calcite precipitation from seawater: I. A classical chemical kinetics description for strong electrolyte solutions. Geochimica et Cosmochimica Acta, 1994. 58(20): p. 4353-4362.

40.   Zuddas, P. and A. Mucci, Kinetics of calcite precipitation from seawater: II the influence of the ionic strength. Geochimica et Cosmochimica Acta, 1998. 62: p. 757-766.

41.   Zuddas, P., K. Pachana, and D. Faivre, The influence of dissolved humic acids on the kinetics of calcite precipitation from seawater solutions. Chemical Geology, 2003. 201: p. 91– 101.

42.   Elfil, H. and H. Roques, Role of hydrate phases of calcium carbonate on the scaling phenomenon. Desalination, 2001. 137: p. 177-186.

43.   Elfil, H. and H. Roques, Prediction of limit of metastable zone in the $CaCO_3$-CO_2-H_2O system. AIChE journal, 2004. 50(8): p. 1908-1916.

44.   Gal, J., et al., Calcium carbonate solubility: a reappraisal of scale formation and inhibition. Talanta, 1996. 43: p. 1497-1509.

45.   Gal, J., Y. Fovet, and N. Gache, Mechanisms of scale formation and carbon dioxide partial pressure influence.Part II. Application in the study of mineral waters of reference. Water Research, 2002. 36: p. 764–773.

46.   Jansen, H., Modelling the Marine Carbonate Pump and its Implications on the Atmospheric CO_2 Concentration. 2001, University Bremen: Bremen.

47.   El-Manharawy, S. and A. Hafez, Could seawater be under saturation and acidic? Desalination, 2004. 165: p. 43-69.

48.   Ingle, S.E., Solubility of calcite in the ocean. Marine Chemistry, 1975. 3: p. 301-319.

49.   Morales, J.G., et al., Precipitation of calcium carbonate from solutions with varying Ca^{2+}/carbonate ratios. Journal of Crystal Growth 1996. 166: p. 1020-1026.

50.   Mullin, J.W., N. Teodossiev, and O. Sohnel, Potassium Sulphate Precipitation from Aqueous Solution by Salting-out with Acetone. Chem. Eng. Process, 1989. 26: p. 93-99.

51.   Dalas, E., J. Kallitsis, and P. Koutsoukos, The crystallization of calcium carbonate on polymeric substrates. Journal of Crystal Growth, 1988. 89: p. 287-294.

52.   Yu, H., The mechanism of composite fouling in Australian sugar mill evaporators by calcium oxalate and amorphous silica, in Chemical Engineering and Industrial chemistry. 2003, University of New South Wales: Sydney.

53.   Treccani, L., Protein Mineral Interaction of Purified Nacre Proteins with Carbonate Crystals. 2006, University of Bremen: Bremen.

54.   Koutsoukoas, P. and C. Kontoyannis, Precipitation of calcium carbonate in aqueous solutions. J. Chem. Soc., Faraday Trans.I, 1984. 80: p. 1181-1192.

55.   Turek, M. and P. Dydo, Electrodialysis reversal of calcium sulphate and calcium carbonate supersaturated solution. Desalination, 2003. 158: p. 91-94.

56.   Turek, M., P. Dydo, and J. Waś, Electrodialysis reversal in high CaSO₄ supersaturation mode. Desalination 2006. 198 p. 288-294.

57.   Dyer, S.J. and G.M. Graham, The effect of temperature and pressure on oilfield scale formation. Journal of Petroleum Science and Engineering 2002. 35: p. 95-107.

58.   Butt, F.H., F. Rahman, and U. Baduruthamal, Hollow fine fiber vs. spiral-wound reverse osmosis desalination membranes Part 2: Membrane autopsy. Desalination 1997. 109: p. 83-94.

59.   Clarkson, J.R., T.J. Price, and C.J. Adams, Role of Metastable Phases in the Spontaneous Precipitation of Calcium Carbonate. J. CHEM. SOC. FARADAY TRANS., 1992. 88(2): p. 243-249.

60.   Dalas, E., P. Klepetsanis, and P.G. Koutsoukos, The Overgrowth of Calcium Carbonate on Poly(vinylchloride-co-vinyl acetate-co-maleic acid). Langmuir, 1999. 15: p. 8322-8327.

61.   Dalas, E. and S. Koklas, The overgrowth of vaterite on functionalized styrene-butadiene copolymer. Journal of Crystal Growth, 2003(256): p. 401-406.

62.   Manoli, F. and E. Dalas, Spontaneous precipitation of calcium carbonate in the presence of ethanol, isopropanol and diethylene glycol. Journal of Crystal Growth, 2000. 218: p. 359-364.

63.   Manoli, F., et al., The effect of aminoacids on the crystal growth of calcium carbonate. Journal of Crystal Growth 2002. 236: p. 363-370.

64.   Kralj, D., L. Bre-eviéa, and A.E. Nielsen, Vaterite growth and dissolution in aqueous solution II. Kinetics of dissolution. Journal of Crystal Growth, 1994. 143 p. 269-276.

65.   Koutsoukos, P.G. and C.G. Kontoyannis, Prevention and inhibition of calcium carbonate scale. Journal of Crystal Growth, 1984. 69· p. 367-376.

66.   Wu, W. and G.H. Nancollas, The dissolution and growth of sparingly soluble inorganic salts: A kinetics and surface energy approach. Pure &Appl. Chem., 1998. 70(10): p. 1867-1872.

67.   Wu, W. and G.H. Nancollas, Determination of interfacial tension from crystallization and dissolution data: a comparison with other methods. Advances in Colloid and Interface Science, 1999. 79: p. 229-279.

68.   Sabbides, T.G. and P.G. Koutsoukos, The crystallization of calcium carbonate in artificial seawater; role of the substrate. Journal of Crystal Growth 1993. 133: p. 13-22.

Chapter 5

Predicting and measurement of pH of seawater reverse osmosis concentrates

Based on parts of:

Waly, T.; Kennedy, M.D.; Witkamp, G.J.; Amy, G. and Schippers, J.C. Predicting CaCO$_3$ scaling: Towards a correct pH calculation in SWRO concentrates. Desalination, submitted

Waly, T.; Kennedy, M.D.; Witkamp, G.J.; Amy, G. and Schippers, J.C. pH calculation in SWRO concentrates. Desalination and water treatment, submitted

Waly, T.; Kennedy, M.D.; Witkamp, G.J.; Amy, G. and Schippers, J.C. Predicting CaCO$_3$ scaling: Towards a correct pH calculation in SWRO concentrates. In Proceedings of the MDIW membranes in drinking and industrial water treatment world congress. 2010. Trondheim, Norway

Munoz, R., Schippers, J.C.; Amy, G. Kennedy, M.D.; Waly, T. and Witkamp, G.J Impact of the ionic strength on the estimation of the pH of the concentrate in Sea water reverse osmosis (SWRO). In proceeding of the 2nd International congress on Water Management in the Mining Industry. 2010. Santiago, Chile

5.1 Abstract

The pH of seawater reverse osmosis plants (SWRO) is the most influential parameter in determining the degree of supersaturation of $CaCO_3$ in the concentrate stream. Accurate determination of the concentrate pH, and in turn its saturation, will play a decisive role in the chemical doses required and hence in the operational costs of any SWRO plant. The acidity constants, ionic strength, temperature, ionic interaction and the rejection of CO_3^{2-}, HCO_3^- and CO_2 by the membranes will also play a role in the resultant concentrate pH. For this calculations making use of different methods were performed and compared with the results of pH measurements of the concentrate of a seawater reverse osmosis pilot plant operating using water of the North Sea. Eight different cases were taken into account based on the CO_2-HCO_3^--CO_3^{2-} system equilibrium equations. In addition, the pH values of the concentrates were calculated using two commercial software programs from membrane suppliers and also the software package Phreeqc. The results of these cases predictions were compared to the concentrate pH measurements in the SWRO pilot plant. Results showed that none of the applied methods was able to predict the measured pH of the concentrate accurately for feed water with pH ranges between 6.6 to 8.5.

In case the pH of the feed water was higher than 7.09 the pH of the concentrate was lower than the pH in the feed water but higher if the feed pH is below 7.09. The selected method assuming that carbon dioxide is not rejected and making use of the first dissociation constant predicted systematically that the concentrate pH will be higher than the measured values. On the contrary, the method based on the second dissociation constant predicted that the concentrate pH value is lower than the measured values.

Furthermore, results suggest that at the lower pH range, the equilibrium between hydrogen carbonate/carbon dioxide dominates the pH in the concentrate while at the higher pH levels the equilibrium hydrogen carbonate/ carbonate dominate. In addition, the effect of using different rejection values for HCO_3^- and CO_3^{2-} showed a marginal influence on the concentrate pH determination compared to the effect of incorporating the acidity constant calculated for NaCl medium or seawater medium.

The outcome of this study indicated that the saturation level of the concentrate was lower than previously anticipated. This was confirmed by shutting down the acid and the antiscalants dosing without any signs of scaling over a period of 6 months. These results confirm that either the concentrate is undersaturated with respect to calcium carbonate or it has a very slow precipitation kinetics.

Keywords: Scaling, equilibrium equations, desalination, membranes, modelling

5.2   Introduction

In seawater reverse osmosis (SWRO) plants where 25-50% of the seawater is converted into fresh water, scaling is a potential hazard for plant designers and operators. It is common practice in seawater reverse osmosis to dose acid or a combination of acid and antiscalant to avoid precipitation of calcium carbonate [1]. The dose of these chemicals is based on the degree of super saturation of calcium carbonate in the concentrate. For the determination of the degree of super saturation the Stiff and Davis Saturation Index (S&DSI) is commonly applied. The outcome of calculating this index results, in almost all cases, in a value which indicates that the concentrate is supersaturated with calcium carbonate.

However, there is some doubt about the need for the addition of acid and/or antiscalant, since some plants don't apply these chemicals at all without being confronted with any scaling of sparingly soluble inorganic compounds i.e. calcium carbonate. Reasons for this situation might be:

- S&DSI does not correctly determine the degree of super or under saturation [2];
- rate of precipitation of calcium carbonate is very slow;
- pH in the concentrate is not correctly predicted.

The aim of this study is to verify the correctness of the pH calculations as in programs which are commonly applied. For this purpose, pH levels in concentrates have been calculated by making use of:

- programs made available by membrane manufacturers;
- manual calculations , making use of dissociation constants of carbon dioxide ,bicarbonate, and carbonate system, which are corrected for salinity;
- Phreeqc program making use of the principle of ion activities.

To verify the outcome of the theoretical calculations, the pH of the feed water and the corresponding concentrate, have been measured in a pilot plant operated with seawater from the North Sea.

5.3   Background

Saturation Indices:

The concept of Saturation Index (SI) is commonly applied to determining whether the concentrate of SWRO plants is supersaturated, in equilibrium or under saturated with calcium carbonate. This SI is by definition the logarithm of the ratio of the product of the concentration of the ionic species involved and the solubility constant. The SI for calcium carbonate is:

$$SI = Log\frac{[Ca^{2+}][CO_3^{2-}]}{K_S}$$ 1.

Where: K_s is the solubility constant.
Langelier [3] derived from this equation the Langelier Saturation Index, which demonstrates the influence of pH.

$$LSI = pH - pH_s$$ 2.

Where:
pH is the real pH.
pH_s is the pH when the water is equilibrium (just saturated) with calcium carbonate.

$$pH_s = pCa + pHCO_3 + pK_{a2} - pK_{sp} = pCa + pHCO_3 + k$$ 3.

Where pCa and $pHCO_3$ are the $-log_{10}$ of calcium and bicarbonate concentration, respectively, and k is a constant depends on temperature and salinity and is obtained graphically [4] or can be calculated based on equation 3 as a function of the second acidity constant and the solubility product of $CaCO_3$ [5]. pK_{a2} is the dissociation constant of HCO_3

Langelier determined the effect of temperature and salinity up to 5000 mg/L on K. Later on, Stiff and Davis [4] expanded the correction for salinity (expressed as ionic strength) to 3.6 mol/L. This index is named Stiff & Davis Stability Index (S&DSI). Langelier and Stiff and Davis corrected the K value for the effect of temperature and salinity on Ka_2 and K_s.

In 1921 Lewis and Randall introduced the concept of activity to explain the effect of temperature and the presence of other ions (salinity/ionic strength) on deviations from law of mass action.

$$a_c = \gamma_i \frac{[C_i]}{[C_\theta]}$$ 4.

a_i is activity of ion i, γ_i is activity coefficient of ion i, $[C_i]$ is concentration of ion i, and $[C_\theta]$ represents the concentration of the chosen standard state, e.g., 1 mol/kg if the work in molality.

In this theory the dissociation and solubility constants are independent from salinity/ionic strength.

Debye and Huckel [6-7] developed a theory enabling to calculate accurately the activity coefficients in solutions up to 0.01 M. Later on Bromley expanded this theoretical concept with the aim to calculate activity coefficients in solutions up to 6 M. The Pitzer specific ion interaction model expands on the Debye – Huckel theory by taking into account long and short range interactions between ions in solutions in calculation activity coefficients. The Pitzer model is valid up to 6 M as well.

Calculation of pH in SWRO concentrates

In SWRO hydrogen carbonate and carbonate are rejected for more than 90 %. As a consequence the concentration of these ions will increase in the concentrate. The concentration factor (CF) depends on the recovery (R) and the rejection (f).

$$CF = \frac{1 - R(1 - f)}{1 - R}$$ 5.

Carbon dioxide is not rejected by membranes. The pH in feed water and concentrate can be calculated with two different relations which are based on the following dissociations.

$$CO_2(aq) + H_2O \leftrightarrow HCO_3^- + H^+$$ 6.

$$HCO_3^- + H^+ \leftrightarrow CO_3^{2-} + H_2O$$ 7.

Traditionally relations are used in which the dissociation constants depend on salinity.

$$pH = pK_{a1} + log \frac{[HCO_3^-]}{[CO_2]}$$ 8.

$$pH = pK_{a2} + log \frac{[CO_3^{2-}]}{[HCO_3^-]}$$ 9.

Both relations are always valid at the same time.

A more recent approach is based on activities of ions involved, which results into:

$$pH = pK_{a1}^o + log \frac{\gamma_{HCO_3^-}[HCO_3^-]}{\gamma_{CO_2}[CO_2]}$$ 10.

$$pH = pK_{a2}^o + log \frac{\gamma_{CO_3^{2-}}[CO_3^{2-}]}{\gamma_{HCO_3^-}[HCO_3^-]}$$ 11.

Where K_{a1}^0 and K_{a2}^0 are acidity dissociation constants at ionic strength of zero mole/L

The ASTM [8] Standard Practice for Calculation and Adjustment of the Stiff and Davis Stability Index for Reverse Osmosis recommend calculating the pH in the concentrate stream by using the ratio of alkalinity (HCO_3) to free CO_2 in the concentrate. A monograph is available to read of the pH in the concentrate stream.
 A note is added regarding seawater systems: For seawater systems the calculated pH of the concentrate can be 0.1 to 0.2 higher than measured if the feed is above 7.0. This statement indicates that the calculation is not accurate. One reason might be the effect of salinity on K_{a1} has not been taken into account.

An indication that the effect of salinity on K_{a1} can not be ignored comes from Copeland, 1967 [20] who reported that seawater evaporated in evaporation ponds decreased in pH from 8.0 to 7.75, when salinity was doubled. The same results were achieved by Hammer and Parker (1984) [21] in their study of Patience Lake where a direct relation between salinity increase and pH decrease was found. This phenomenon seems to hold when salinity decreases, where Neev and Emery, 1967 [9] and Amit and Bentor, 1969 [10] showed that the pH values of the Dead Sea water increase progressively when it is diluted with distilled water. This pH depression has to be attributed to the relation between the acidity constants and pH in a hyper-saline environment [11-12].

Effect of salinity on dissociation constants

Several researchers investigated the dissociation constants of carbon dioxide and hydrogen carbonate in synthetic (NaCl solutions with the same ionic strength as that of seawater) and real seawater (See Table 1).

Table 1: Range of salinity and temperature covered by various researchers for the determination of the acidity constants in synthetic water (SSW) and real seawater (SW) composition

Reference Name	Temp. range	Salinity range	Type of water
Hansson, 1973 [13]	5-30	20-40	SSW
Meherbach et al, 1973 [14]	5-45	5-43	SW
Goyet et al, 1989 [15]	10-50	1-40	SSW
Roy et al, 1993 [16]	0-45	5-45	SSW
Lee and Millero, 1995 [17]	0-58	33-37	SW
Lueker et al, 2000 [18]	5-45	5-43	SW
Prieto & Millero, 2002 [19]	15-45	5-43	SW
Millero et al 2006 [20]	0-50	1-50	SW

The outcome of these studies is that K_{a1} and K_{a2} in sodium chloride and seawater (I=1 mol/L) are higher than in low salinity water (Table 2).

Table 2: Acidity constants at different ionic strength levels and in seawater medium and NaCl medium at 25°C [6, 20-21]

Constant	Ideal solution	NaCl medium [22]	SW medium [20]
	I=0 mole/L	I=1 mole/L	I=1 mole/L
pK_{a1}	6.35	5.97	5.79
pK_{a2}	10.34	9.48	8.69

Millero et al.[20], developed detailed equations to calculate the dissociation constants from 1 ‰ to 50 ‰ salinity and 0 °C to 50 °C. Their work was based on the extensive studies of Meherbach et al, 1973 [14] to determine the acidity constants on seawater, These salinity and temperature ranges are applicable to almost all natural seawaters but not to all SWRO concentrates where sometimes salinities higher than 50‰ are found in concentrates.

$$pK_{a1}^{TS} = -126.34048 + \frac{6320.813}{T} + 19.568224Ln(T)$$
$$+ 13.4191S^{0.5} + 0.0331S - 5.33*10^{-5}S^2 \qquad \text{12.}$$
$$- \frac{530.123S^{0.5} + 6.103S}{T} + -2.0695S^{0.5}Ln(T)$$

$$pK_{a2}^{TS} = -90.18333 + \frac{5143.692}{T} + 14.613358Ln(T)$$
$$+ 21.0894S^{0.5} + 0.1248S - 3.687*10^{-4}S^2 \qquad \text{13.}$$
$$- \frac{772.483S^{0.5} - 20.051S}{T} + -3.3336S^{0.5}Ln(T)$$

Where K_{a1}^{TS} and K_{a2}^{TS} are the acidity constants in seawater medium, T is the temperature in degree Kelvin and S is the salinity in ‰.

Effect of salinity on the ion activity

Computer software such as Phreeqc enables to calculate ion activities. These calculations were initially based on the Debye-Huckel-Bromley equations. The most recent version makes use of the Pitzer model. The equations used in this model to calculate the ionic activities, are rather complex, which is illustrated below.

$$\ln(\gamma_x) = |z^+|^2 \left\{ -A^* \left(\frac{\sqrt{I}}{1+1.2\sqrt{I}} \right) + \frac{2}{1.2}\ln(1+1.2\sqrt{I}) \right.$$
$$+ \sum_a \sum_c m_c m_a \beta_{ca}^{"}$$
$$\left. + \sum_a \sum_{a"} m_a m_{a"} \emptyset_{aa"}^{"} + \sum_c \sum_{c"} m_c m_{c"} \emptyset_{cc"}^{"} \right\}$$
$$+ \sum_a m_a \left\{ 2\beta_{xa}^{"} \right. \qquad \text{14.}$$
$$+ \frac{1}{2}C_{xa} \left(\sum_c m_c |z^+|_c + \sum_c m_a |z^+|_a \right) \right\}$$
$$+ \sum_a m_a \left\{ 2\emptyset_{xc}^{"} + \sum_a m_a \psi_{xca} \right\}$$
$$+ \sum_a \sum_{a"} m_a m_{a"} \psi_{xaa"} + |z^+| \sum_c \sum_{a"} m_c m_a C_{ca}$$

$$
\begin{aligned}
\ln(\gamma_y) = |z^-|^2 \Bigg\{ &-A^* \left(\frac{\sqrt{I}}{1 + 1.2\sqrt{I}} \right) + \frac{2}{1.2} \ln(1 + 1.2\sqrt{I}) \\
&+ \sum_a \sum_c m_c m_a \beta''_{ca} \\
&+ \sum_a \sum_{a''} m_a m_{a''} \phi''_{aa''} + \sum_c \sum_{c''} m_c m_{c''} \phi''_{cc''} \Bigg\} \\
&+ \sum_c m_c \Bigg\{ 2\beta''_{cy} \\
&+ \frac{1}{2} C_{cy} \left(\sum_c m_c |z|_c + \sum_c m_a |z|_a \right) \Bigg\} \\
&+ \sum_a m_a \Bigg\{ 2\phi_{ya} + \sum_c m_c \psi_{yac} \Bigg\} \\
&+ \sum_c \sum_{c''} m_c m_{c''} \psi''_{yaa''} + |z^+| \sum_c \sum_a m_c m_a C_{ca}
\end{aligned}
$$

15.

Where β'' and C are function of the Pitzer ionic interaction parameters, Φ and Φ'' second virtual coefficient for symmetric and unsymmetrical ions, Ψ and Ψ'' a third virtual mixing parameters in Pitzer equation, x and m cations of interest, y and c anions of interest, z+ cation charge, z- anion charge, Z_{M1} valance of M in MX supersaturated solution, Z_{X1} valance of X in MX supersaturated solution, m is the molar concentration, the summation index, c or a, denotes the sum over all cations or anions in the system while the double summation index, c and c" or a and a", denotes the sum over all distinguishable pairs of dissimilar cations or anions; m is the molality of the species; I is the ionic strength of the solution.

Combining these activities with concentrations and $K^o{}_{a1}$ and $K^o{}_{a2}$ (thermodynamic dissociation constants at 25 °C and I = 0 mol/L) results in the pH according to equations 10 and 11. These thermodynamic dissociation constants are independent from ionic strength (salinity)

5.4   Materials and Methods

The verification of the outcome of the theoretical calculations was carried out in four steps:

- pH prediction using equilibrium equations, taking into account the effect of salinity on dissociation constants
- pH prediction using computer software supplied by membrane manufacturers
- pH prediction with the evaporation model of Phreeqc
- pH measurements in a pilot plant treating North seawater.
- Compeering the results of the theoretical calculations with the data obtained from field measurements;

Pilot plant

All of the work in this research was carried out on real SWRO concentrates from a pilot plant using North Sea water as its feed water and located in the south of the Netherlands. In the pilot plant process scheme shown in Figure 1, the feed water is filtered through a 150μm strainer before the addition of acid to decrease the pH of the feed water from 8.0 to 6.7.. The water is then fed to a UF (X-Flow) membrane of 300 kDa (0.090 μm) filter pore size before being fed to the RO unit. The water recovery of the RO unit is around 40%. The feed water is considered to be undersaturated according to S&DSI (-0.06) and oversaturated when the saturation is calculated using the SI approach by the PhreeqC program (SI=0.42), based on solubility of calcite. In the SI calculations, the program incorporates the Pitzer activity coefficient in the calculations to account for interaction between ions and uses the solubility of calcite in its calculations. Although the plant is designed to have an antiscalant addition, the antiscalant system was not used.

In a normal operational mode, the pre-treated SWRO water feeding the high pressure pump has a pH of 6.7 after acidification.

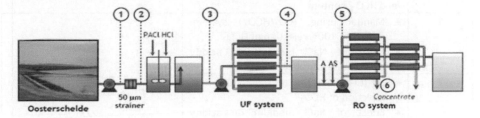

Figure 1: Schematic diagram for the Zeeland pilot plant showing the plant's treatment steps

Table 3: The ionic composition and the saturation of the feed seawater at Zeeland

Ions		Inlet (Summer)	Concentrate (R=40%)	Inlet (Spring)	Concentrate (R=40%)
Temperature	°C	17.5	17.5	10	11.6
HCO_3^-	mg/L	154	256	154	256
Calcium	mg/L	350	583	350	583
Magnesium	mg/L	1100	1833	1100	1833
Sodium	mg/L	9700	16166	9700	16166
Potassium	mg/L	370	616	370	616
Boron	mg/L	4.1	7	4	7
Chloride	mg/L	16900	28166	16900	28166
Sulphate	mg/L	2551	4251	2551	4251
Ammonium	mg/L	0.06	0.1	0.06	0.1
Strontium	mg/L	7.1	12	7.1	12
Silica	mg/L	0.56	0.9	0.6	0.9
TDS	mg/L	31137	51894	31137	51894
Ionic strength	mole/L	0.62	1.03	0.62	1.03

Concentrate pH in SWRO systems

A- pH calculation taking into account the effect of salinity on the dissociation constants

Using the equilibrium equations represented by equations 8 and 9, eight calculation cases were developed. The cases were divided into 2 groups based on the species used in the concentrate pH calculations. In the first group the CO_2-HCO_3^- equilibrium (equation 8 and Table 4) is used in the pH calculations. The main assumption is that CO_2 in the feed stream to the SWRO system is equal to that in the concentrate streams (i.e., no rejection) CO_2 content in the feed water was calculated using the pH and HCO_3^- in the feed water with equation 8.

Table 4: Summary of group 1 for pH prediction calculations

Scenario	Groups for calculation of pH	Equilibrium equation used
	Group 1: Concentrate pH assumes that feed CO_2 is calculated from feed pH and feed HCO_3^- content	
1	• Manual using, (CO_2/HCO_3^-) system, assuming 100% rejection of HCO_3^-. • Effect of NaCl medium on acidity constants is considered	$pH = pK_{a1}^{TNaCl} + log\dfrac{[HCO_3^-]}{[CO_2]}$
2	• pH (CO_2/HCO_3^-) system, assuming 90% rejection of HCO_3^- • Effect of NaCl medium on acidity constants is considered	
3	• Manual using (CO_2/HCO_3^-) system, assuming 100% rejection of HCO_3^-. • Effect of SW medium on acidity constants is considered	$pH = pK_{a1}^{TS} + log\dfrac{[HCO_3^-]}{[CO_2]}$
4	• pH (CO_2/HCO_3^-) system, assuming 90% rejection of HCO_3^-. • Effect of SW medium on acidity constants is considered	

In the second group (Table 5) the CO_3^{2-}-HCO_3^- equilibrium equation (equation 10) was applied in the pH calculation. The amount of feed CO_3^{2-} has been determined by using the feed pH and feed alkalinity in equation 9. The pH of the concentrate can then be calculated by knowing the rejection values of CO_3^{2-} and HCO_3^- in the concentrate and the second acidity constant value in a saline medium corresponding to the same salinity as that of the concentrate.

Table 5: Summary of group 2 for pH prediction calculations

Scenario	Groups for calculation of pH	Equilibrium equation used
	Group 2: Concentrate pH assumes that feed CO_3^{2-} is calculated from feed pH and feed HCO_3^- content	
5	• Manual using (CO_3^{2-}/HCO_3^-) system, assuming 100% rejection of HCO_3^-. • Effect of NaCl medium on acidity constants is considered	$pH = pK_{a1}^{TNaCl} + log \dfrac{[CO_3^{2-}]}{[HCO_3^-]}$
6	• pH (CO_3^{2-}/HCO_3^-) system, assuming 90% rejection of HCO_3^- and 99% of CO_3^{2-} • Effect of NaCl medium on acidity constants is considered	$pH = pK_{a1}^{TS} + log \dfrac{[CO_3^{2-}]}{[HCO_3^-]}$
7	• Manual using (CO_3^{2-}/HCO_3^-) system, assuming 100% rejection of HCO_3^- and CO_3^{2-}. • Effect of SW medium on acidity constants is consider	
8	• pH (CO_3^{2-}/HCO_3^-) system, assuming 90% rejection of HCO_3^- and 99 % of CO_3^{2-}. • Effect of SW medium on acidity constants is consider	

Two of the four cases were dedicated to taking into account the effect of salinity on the calculated concentrate pH. The increase in the salinity affects the acidity constants and hence, the calculated concentrate pH. The difference between seawater medium and NaCl medium that the pH was calculated using K_{a1}^{TS} and K_{a2}^{TS} in the cases simulating the ionic interaction in seawater versus K_{a1}^{TNaCl} and K_{a2}^{TNaCl} in the cases simulating NaCl medium. K_{a1}^{TS} and K_{a2}^{IS} were calculated based on equation 12-13 while K_{a1}^{TNaCl} and K_{a2}^{TNaCl} were calculated using equations 16-17 developed by Millero et al, 2007 [23] for acidity constants in NaCl medium from 0-50°C and up to ionic strength of 6 moles/L, and were verified with values derived from the literature data as shown in Table 6.

Table 6: Values of first and second acidity constants in a NaCl solution at 17.5°C

Ionic strength mole/L	0	0.62	1.03	Adopted & calculated after
pK_{a1}^{NaCl}	6.3995	6.0198	5.991	[24] & [25]
pK_{a2}^{NaCl}	10.405	9.626	9.556	[26] & [22]

$$pK_{a1}^{TNaCl} == -114.3106 + \frac{5773.67}{T} + 17.779524Ln(T)$$
$$+ 35.2911 I^{0.5} + 0.8491 I - 0.32 I^{1.5} + 0.055 I^2 \qquad \text{16.}$$
$$- \frac{1583.09 I^{0.5}}{T} - 5.4366 I^{0.5}{}_1 Ln(T)$$

$$pK_{a2}^{TNaCl} == -83.2997 + \frac{4821.38}{T} + 13.5962Ln(T) + 8.2746I^{0.5}$$
$$+ 1.6057I - 0.6471I^{1.5} + 0.1131I^2 - \frac{1738.16I^{0.5}}{T} \qquad 17.$$
$$- 6.0346I^{0.5}Ln(T)$$

Where $K_{a1}{}^{TNaCl}$ and $K_{a2}{}^{TNaCl}$ are the acidity dissociation constants in NaCl medium, T is the temperature in degree Kelvin and I is the ionic strength in mole/L.

The other two cases were used to simulate different membrane rejection ratios for HCO_3^- and CO_3^{2-} ions. It is well known that the membrane rejection of monovalent ions is lower than that of divalent ions resulting in more HCO_3^- passing through the membrane than CO_3^{2-}. In these specific cases, rejection figures of 90% and 99% were assumed for the HCO_3^- and CO_3^{2-}, respectively, and compared to 100% membrane rejection to explore the effect of rejection differences on pH calculations. The concentration factor was calculated based on equation 5 for the two scenarios applied (Table 7).

Table 7: Concentration factor for carbonate and bicarbonate taking into account different membrane rejection factors

HCO₃⁻				CO₃²⁻		
CF	Recovery	Salt rejection		CF	Recovery	Salt rejection
1.67	40%	100%		1.67	40%	100%
1.60	40%	90%		1.66	40%	99%

B- pH prediction using membrane manufacturers programs

In this part of the research, two membrane manufacturers software, namely, Program 1 and Program 2, were used to access the pH prediction cases developed. These programs are free web programs used to design SWRO plants and developed by two membrane manufacturer companies. The two programs provide estimates of the pH of the concentrate as well as the membrane's ionic rejections. Based on the calculated pH and consequently the expected concentrate saturation, the computer programs recommend whether or not antiscalants or acids are needed. The two programs use to give a warning messages when the S&DSI index exceeds zero. The recommended chemical doses are designed to maintain the S&DSI value in the concentrate equal to zero. When no chemicals are used warning statements are written on the final printout indicating that the concentrate water is precipitative. This software allows an assessment of the RO configuration specifying the RO membrane type, number of pressure vessels and elements per pressure vessels, for a given feed flow and recovery, although this tool was not used in the simulation since the only required output was the scaling part.

C- pH prediction using Phreeqc evaporation model

The Phreeqc software is an equilibrium program which has been used mainly for groundwater modelling. The program uses Pitzer activity coefficients to account for the ionic strength and ionic interaction in the concentrate. Using Pitzer activity coefficients in the calculations, allows not only the ion pairs formation to be taken into consideration but also the ionic interaction between same charged ions in seawater samples. The simulation of the SWRO process in Phreeqc can be carried out using the Evaporation Model incorporated in the program, as the program assumes both a decrease in water content due to evaporation and also concentration of all the salt in the water that remains after evaporation. The program computes the new ionic composition and the pH of the concentrate with the option to prevent CO_2 exchange with atmosphere [27]. The main difference between the evaporation model and real situation in membrane elements, is that in the evaporation model the total inorganic carbon is concentrated leading to an increase in the concentrate CO_2, while in membrane element, the CO_2 is not rejected. Nevertheless, the evaporation model can be used as an indication for the concentrate pH especially when the CO_2 in the feed stream is minimal.

D- pH measurements of feed water and concentrate

The pH of the seawater and the concentrate were measured in a pilot plant. For this purpose the SWRO system was stopped and flushed with un-chlorinated product water. When re-operating the plant, the acid dose was stopped for 2 hours before collecting the concentrate samples. The pH was measured using a highly sensitive offline pH meter (Eutech 6000) capable of measuring with three-digit pH accuracy up to an ionic strength of 1.6 mole/L. The feed water pH was 8.0 - 8.02 and was monitored using online pH meters as well (Endress & Hauser) located on the feed stream of the UF and on the feed stream to the high pressure pump. A feed sample was taken after the micro-strainer for offline pH measurements to insure the integrity of the online measurements.

The concentrate sampling was carried out by taking five samples from the SWRO concentrate stream in plastic clean vessels. In order to minimize gas exchange with atmospheric pressure, the flow rate of the concentrate to the vessels was kept as low as possible (1 L/min). The vessels were completely filled and sealed from atmospheric air before being transported to the lab for pH measurement.

In order to expand the range of pH values in the feed water and concentrate, there was an extended campaign in the spring time (2010) during the algal bloom period to experience the maximum feed water pH of 8.5. The pH was varied from 8.5 in steps to 6.6 by means of acid dosing. The plant was kept running for 2 hours for pH stabilization after each change in the feed pH after which the feed to the RO and the concentrate were collected and the pH was measured.

5.5   Results and Discussion

pH calculation using equilibrium equations

The difference in the equilibrium equation used in the pH calculation affects the expected concentrate pH, as shown in Table 8 and Figure 2. Results from the first group using the CO_2-HCO_3^- equilibrium equation, showed that the CO_2 content in the feed equals $2.13*10^{-5}$ for NaCl medium and $2.64*10^{-5}$ mole/L for seawater medium. The results suggest that concentrate pH will increase compared to the feed (8.0); the highest expected pH value was in the range of 8.19-8.18 when NaCl medium is taken into consideration and 8.15-8.13 when seawater medium is considered. The highest value of 8.19 was found in NaCl medium when 100% rejection was considered for all ions. In contrast, the lowest pH value of 8.13 was found when seawater medium and lower rejection values were applied. The effect of using K_{a1}^{TNaCl} and K_{a1}^{TS} was 0.04 pH units which is 4 times the change in pH due to applying different rejection ratios (0.01 pH units). This small decrease can be attributed to the minor difference between the first acidity constant values used (K_{a1}^{TNaCl} =5.99 and K_{a1}^{TS} =5.86) for ionic seawater medium and NaCl medium.

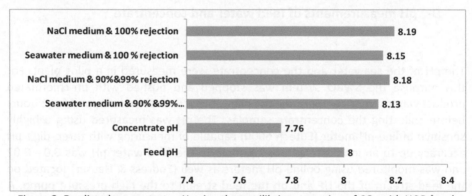

Figure 2: Predicted concentrate pH using the equilibrium equation of CO_2 with HCO_3^- and assuming that the value of CO_2 in feed water is equal to that using feed pH and CO_2-HCO_3^- equilibrium

Figure 3 represents the predicted pH of the second group utilizing the CO_3^{2-}-HCO_3^- equilibrium equation (equation 10). Results show that the calculated concentrate pH was lower than that calculated from the first group and even lower than the feed pH (Figure 3). The pH values for the concentrate varied between 7.95 (case 6) and 7.74 (case 7). The first represents pH values when NaCl ionic interaction and only different rejection ratios were considered while the latter considered the effect of seawater medium and ignored the effect of rejection ratios. The effect of changing the rejection values on the predicted pH was similar to that found in the first group (0.02 pH units). The different rejection ratios used and its minor effect suggest that using 100% rejection in SWRO calculations will not greatly affect our expected outcome and calculations.

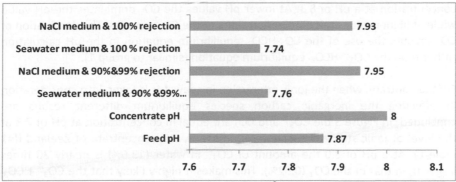

Figure 3: Predicted concentrate pH using the CO_3^{2-} - HCO_3^- equilibrium equation

The results in the second group suggest that the concentrate pH prediction is largely affected by using the CO_3^{2-}-HCO_3^- equilibrium (equation 10) compared to the CO_2-HCO_3^- (equation 9). This may be attributed to two reasons; the first is related to the increase in both ions (CO_3^{2-}-HCO_3^-) in the concentrate due to membrane rejection, contrary to group 1 where the CO_2 was always kept constant in the feed and the concentrate. The second reason is related to the large decrease in the second acidity constant due to the increase in salinity if compared to the decrease in the first acidity constant used in group 1. Cases 7 and 8 showed a larger decrease in the predicted pH (7.74 and 7.76) if compared to cases 5 and 6 (7.93, 7.95), where the acidity constants were calculated for NaCl medium. These results in the second group match the findings of Copeland, 1967 [28] who recorded a decrease of pH when seawater is concentrated by evaporation.

Table 8: Group distribution and scenario developments

Case	pH	I(mole /l)	TDS (mg/L)	pK_{a1}	pK_{a2}	HCO_3^- (mole/L)	CO_3^{2-} (mole/L)	CO_2(mole /L)
Feed NaCl medium	8.00	0.62	31137	6.02	9.63	2.52E-03	5.96E-05	2.64E-05
Feed Seawater medium	8.00	0.62	31137	5.93	9.05	2.52E-03	2.26E-04	2.13E-05
Group 1								
1	8.18	1.03	51894	5.99	9.56	4.08E-03		2.64E-05
2	8.18	1.03	51894	5.99	9.56	3.91E-03		2.64E-05
3	8.15	1.03	51894	5.86	8.80	4.08E-03		2.13E-05
4	8.13	1.03	51894	5.86	8.80	3.91E-03		2.13E-05
Group 2								
5	7.93	1.03	51894	5.99	9.56	4.08E-03	9.66E-05	
6	7.95	1.03	51894	5.99	9.56	3.91E-03	9.58E-05	
7	7.74	1.03	51894	5.86	8.80	4.08E-03	3.66E-04	
8	7.76	1.03	51894	5.86	8.80	3.91E-03	3.63E-04	

Figures 4 and 5 represent the effect of ionic interaction in seawater medium on the inorganic carbon species equilibrium. In Figure 4 the ionic interaction was ignored and values of the acidity constants for I=0 were used (Table 6). Results showed that at an ionic strength of zero, the CO_3^{2-} and CO_2 species in water are equal in concentration at a pH of 8.3. At lower pH values the CO_2 dominates the pH value while at higher values this domination shifts to the CO_3^{2-} species. The domination of CO_2 favours the use of the CO_2-HCO_3^- equilibrium equation in the pH calculation rather than the CO_3^{2-}-HCO_3^- equilibrium equation (similar to group 1).

On the contrary, when the ionic interaction in seawater is taken into consideration in affecting the inorganic carbon species equilibrium different results are concluded. In Figure 5 the CO_3^{2-} and CO_2 are equal in concentration at pH of 7.3 at the level of ionic strength encountered in the SWRO concentrate of Zeeland (I=1 mole/L). At a pH of 8.0 the amount of CO_3^{2-} in water (13.6%) is nearly 20 times larger than that of the CO_2 (0.62%). This makes it highly likely that the CO_3^{2-}-HCO_3^- predominates the pH calculation and not the CO_2-HCO_3^-. This may suggest that the pH of the concentrate in real SWRO systems will follow the same trend of decrease with increasing salinity found in group 2 (Table 8). This hypothesis still needs further confirmation.

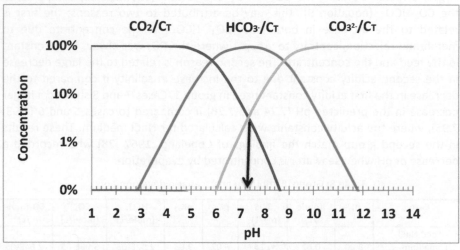

Figure 4: Relation between different inorganic carbon species with respect to solution pH for SWRO concentrate concentration of 50,000 mg/L (I=1.0) at 17.5°C.

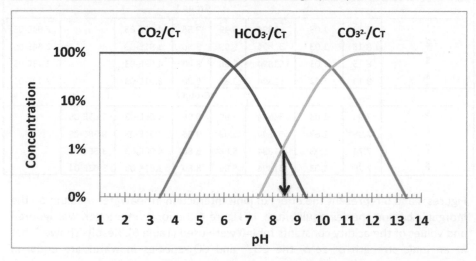

Figure 5: Relation between different inorganic carbon species and total inorganic carbon (C_T) at zero ionic strength with respect to solution pH at $17.5^{\circ}C$.

pH calculations using membrane manufacturers software

The results shown in Table 9 and Figure 6 for different software calculations show large differences In the predicted concentrate pH among these programs. The highest pH value of 8.5 was calculated by program 1 while the lowest pH of 7.92 was calculated by program 2 final report. Program 1 depends mainly on the assumption that the CO_2 in the feed is equivalent to that of the concentrate and equals 0.81 mg/L. When calculating the acidity constants using the data in program 1 final report it showed, contrary to previous calculations, an increase in the pK_{a1} and pK_{a2} with the salinity increase resulting in considerable pH increase in the concentrate. Similar results were found in the literature when the activity coefficients are calculated using the Davis equation in water with an ionic strength larger than 0.5 mole/L. In such cases a sharp decrease is experienced in the pK_{a1} and pK_{a2} up to I = 0.5 mole/L followed by a smoother increase when the ionic strength of the solution is increased [21]. Never the less, it is worth mentioning that Davis equation is only valid up to ionic strength of 0.5 mole/L .

Table 9: The calculation of the first and second acidity constants used by Program 1 and Program 2

	Program 1		Program 2 scaling calculation		Program 2 final report	
	feed	Conc.	feed	Conc.	feed	Conc.
pH	8	8.5	8	8.21	8	7.92
CO_2 (mg/L)	0.81	0.81	0.94	0.94	0.94	1.70
CO_3^{2-} (mg/L)	11	24	13	22	13	24
HCO_3^- (mg/L)	154	254	154	256	154	252
pK_{a1}	6.13	6.42	6.12	6.12	6.12	6.10
pK_{a2}	9.13	9.63	9.05	9.26	9.05	8.93

In program 2, two predicted concentrate pH values of 8.21 and 7.92 were reported by the program; the first pH was used by the program for the scaling assessment and chemical dose recommendations for the feed water. The program used the same assumptions used by program 1 to calculate the pH where the feed and the concentrate CO_2 content were kept constant and equal to 0.937 mg/L. Using the data provided by the program, the calculation shows that in such cases the pK_{a1} value for the feed and the concentrate were kept constant and pK_{a2} was slightly higher in the concentrate than that of the feed (9.26 and 9.05, respectively).

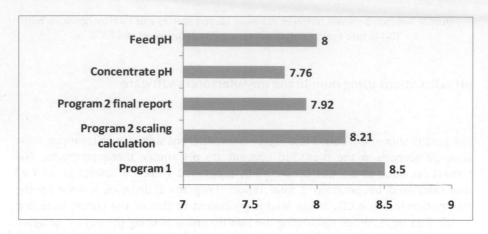

Figure 6: Predicted concentrate pH using program 1 and program 2 compared to real values measured for the SWRO plant concentrate

Although there were no chemicals added to justify the feed pH and the warning signs for scaling threats had disappeared, the final program 2 report showed a much lower expected concentrate pH of 7.92 (instead of 8.21). The calculation of the acidity constants using the data of program 2 (Table 9), showed the acidity constants values for the feed water were not changed by the program when performing the scaling calculations for the concentrate (6.12 and 9.05). On the other hand, the acidity constants for the concentrate at the final report were lower compared to the feed water especially, for the second acidity constant (8.93). These results suggest that the program may be using the CO_2-HCO_3^- system equilibrium equation in the calculation of the scaling tendency of the concentrate while using the CO_3^{2-}-HCO_3^- equilibrium equation in the concentrate pH appeared on the final report. Although the program assumes the feed and the concentrate CO_2 content are the same in the feed and the concentrate, in the final report an increase of 81% in the CO_2 content in the concentrate was reported. This increase is either attributed to partial rejection of the dissolved CO_2 by the membrane or due to the increase in CO_3^{2-} and HCO_3^- content in water which in turn affects the amount of CO_2 in the concentrate.

pH prediction using Phreeqc evaporation model

Table 10: Calculation of the first and second acidity constants calculated after Phreeqc when using the Pitzer activity coefficients for seawater and NaCl mediums.

	Phreeqc (Pitzer) SW		Phreeqc (Pitzer) NaCl	
	feed	Conc.	feed	Conc.
pH	8.0	7.86	8.0	7.91

pH calculations were done using the Phreeqc evaporation model, integrating the ionic complexation in the calculation of the activities using Pitzer activity coefficients. In seawater medium, the predicted pH calculated by Phreeqc was 7.86, which is lower than pH values predicted in groups 1 and 2. The pH value in seawater medium is lower by 0.04 pH units compared to that obtained in NaCl medium (Figure 7). In NaCl medium, the predicted pH value was 7.91 which is very similar to that obtained in case 5 when NaCl ionic interaction is taken into consideration.

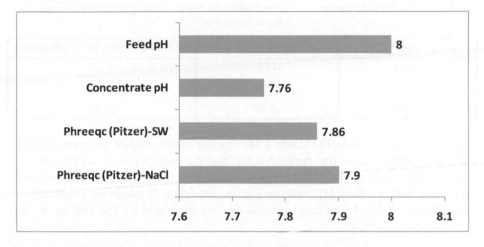

Figure 7: Predicted concentrate pH using Phreeqc compared to real values measured for the SWRO plant concentrate

Field seawater concentrate measurements:

A- Comparing theoretical calculations with field measurements at pH 8 of the feed water

When the pH of the feed water in Zeeland SWRO pilot plant was 8, the pH of concentrate (Table 11 and Figure 7) showed an average pH value of 7.76±0.02 (5 samples) and S&DSI of 0.07. These values are lower than all of the calculated values in the 8 cases as well as the concentrate pH values calculated by the commercial software and with the PhreeqC evaporation model. The nearest value in the 8 cases was pH = 7.74 and 7.76 belonging to scenarios 7 and 8 which took into account the effect of seawater ionic interaction on the CO_3^{2-}-HCO_3^- system. The similarity in values between the field and calculations suggests confirmation of the correction of the acidity constants used in this research. This decrease in concentrate pH due to the increase in salinity is similar to what was found earlier by oceanographic researchers [9-12, 28] and modelling results in this research. This may suggest that the scaling tendency of SWRO concentrates may not be as severe as was previously assumed.

The software program that predicted pH values that were closest to real measurements was the Phreeqc (Pitzer) evaporation model (pH = 7.86) which is higher by nearly 0.1 pH units than the real measurement. This may be attributed to the fact that Pitzer constants, some organic and inorganic complexes that may be formed in real seawater are not taking into account and hence might decreasing the overall available amount of free HCO_3^- and CO_3^{2-} in water and affecting the acidity constants [19-20].

Table 11: The feed and concentrate pH and S&DSI for SWRO working at 40% recovery using North Sea water

Ions	Inlet without acidification	SWRO Conc. 40% (no acidification)
Temp	17.5	17.5
pH	8.00	7.76
S&DSI	-0.06	+0.07

The SWRO real concentrate showed a S&DSI value slightly higher than zero (+0.07). This may suggest that the concentrate is slightly oversaturated with regard to $CaCO_3$. On the other hand, it is worth mentioning that the S&DSI index is developed only in a NaCl medium [4], and the k value (equation 3) obtained graphically is expected to differ if measured in seawater medium due to the change in the second acidity constant.

It is worth mentioning that the pilot plant has been working at pH of 8.0 for several months without the addition of acids or antiscalants. These results suggest that for SWRO systems working at a recovery rate of 40% and using seawater of the North Sea with feed pH of 8.0, the concentrate is undersaturated with respect to $CaCO_3$ And/or having slow precipitation kinetics of $CaCO_3$.

B- Comparing theoretical calculations with field measurements at pH range of 6.6 to 8.5 of the feed water

For the purpose of comparing approach followed in scenario 3 and 7, 15 measurements for the feed and the concentrate were carried out at different feed pH levels (Table 12 and Figure 8). The results indicate that the concentrate pH is lower than the feed pH when the feed pH is higher than 7.09. On the contrary, the concentrate pH is higher than that of the feed, when the feed pH is lower than 7.09.

The results represented in Figure 8 indicate that the field concentrate pH measurements did not match either pH values predicted using the CO_2-HCO_3^- or CO_3^{2-}-HCO_3 equilibrium equations. These results suggest that at feed pH values lower than 7.09 the CO_2-HCO_3^- equilibrium equation is more influential in determining the concentrate pH over CO_3^{2-}-HCO_3 equilibrium equation, forcing the resultant pH in the concentrate to increase. On the other hand, at feed pH values higher than 7.09, the CO_3^{2-}-HCO_3 equilibrium equation seems to be more influential and the concentrate pH tends to decrease compared to the feed pH. Furthermore, at pH of 7.09, the two equilibrium equations appear to have equal influence on determining the concentrate pH with the no change in the concentrate pH compared to the feed.

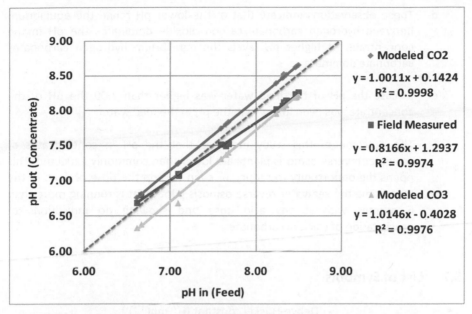

Figure 8: The concentrate pH vs. feed pH for field measurements and predicted pH using equilibrium equations. The dotted line is a 45° line (line where the pH of the feed equals the pH of the concentrate).

The relation between the field measured and the predicted pH suggest a straight line with an R^2 value greater than 0.997 (equations 36-38). These developed fittings equations suggests that the pH predicted using the CO_3^{2-}/HCO_3 equilibrium equation should match that of the measured value at a feed pH of 8.85 (full dominance of the CO_3^{2-}/HCO_3^- on concentrate pH). On the other hand, the predicted pH using the CO_2/HCO_3^- equilibrium equation will match that of the field measurement at a pH of 6.32 (full dominance of the CO_2/HCO_3^- in the pH determination).

5.6   Conclusions

a- None of the applied methods predicted the pH of the concentrate accurately in the range of pH 6.6 to 8.5 of the feed water.

b- The selected method assuming that carbon dioxide is not rejected and making use of the first dissociation constant predicted systematically pH value higher than the real pH. The lower the pH in the feed water the closer prediction to the measured pH.

c- The method based on the second dissociation constant predicted systematically pH value lower than the real pH. The higher the pH the closer the prediction to the measured pH.

d- These observations indicate that at the lower pH range the equilibrium between hydrogen carbonate/carbon dioxide dominates the pH in the concentrate. At higher pH levels the equilibrium hydrogen carbonate/ carbonate dominate.

e- In case the pH of the feed water was higher than 7.09 the pH of the concentrate was even lower than the pH in the feed water.

f- The outcome of this study indicates that the pH in concentrates of seawater reverse osmosis plants are lower than commonly expected. This opens the opportunity to reduce or even to stop the dose of acid. In the mean time the seawater reverse osmosis pilot plant is running more than 6 months without any acid dose and showing no indications of precipitation of calcium carbonate.

5.7 List of Symbols

$A*$	Debye-Huckel constant ($L^{2/3}mol^{-1/2}$)
a	Ionic effective diameter (m)
a_c	Activity
Alk	Alkalinity (mole/L)
B	Constant in extended Debye-Huckel equation ($L^{-2/3}mol^{1/2}.m$)
b	Constant depends on the method of activity calculation ($L^3 mol^{-1}$)
C	In equations 40, 41 is function of the Pitzer ionic interaction parameters
C_T	Carbon total (mole/L)
f	Salt rejection ratio on membrane surface
f_{CO_2}	Fugacity of CO_2 (atmospheres)
I	the ionic strength (mol/L)
K_{so}^{o}	Solubility at standard conditions ($mole^2/L^2$)
K_{a1}^{o}	First acidity constant in pure water (mole/L)
K_{a2}^{o}	Second acidity constant in pure water (mole/L)
K_{a1}^{TNaCl}	First acidity constant temperature and salinity corrected in NaCl medium (mole/L)
K_{a1}^{TNaCl}	Second acidity constant temperature and salinity corrected in NaCl medium (mole/L)
K_{a1}^{TS}	First acidity constant temperature and salinity corrected in seawater medium (mole/L)
K_{a2}^{TS}	Second acidity constant temperature and salinity corrected for seawater medium (mole/L)
K_{H}^{o}	Henry's constant at ambient temperature, ambient pressure and at I = 0 (mole.atm/L)
m	Morality (mole/L)

P_{CO2}	Partial pressure of carbon dioxide (atmosphere)
pH_s	pH at saturation or equilibrium
S	Salinity ‰ is parts per thousands (Weight$_{salts}$ grm/Weight$_{water}$ Kg)
SI	Supersaturation Index
S&DSI	Stiff and Davis Saturation Index
T	Absolute temperature in Kelvin
TCF	Temperature correction factor
TDS	Total dissolved solids in the feed stream (mg/L)
[X]	Molecular weight species X (mole/L)
[X$_o$]	Molecular weight species X in ideal state (mole/L)
X&m	Cation of interest
Y&c	Anion of interest
z_+	Cation charge
z_-	Anion charge
z_{M1}	Valance of M in MX supersaturated solution
z_{X1}	Valance of X in MX supersaturated solution

Greek letters

γ_+	Cation activity coefficient
γ_-	Anion activity coefficient
γ	Mean activity coefficient
γ_X	Activity coefficient of X
$\gamma_{X(NaCl)}$	Activity coefficient of X in a NaCl medium
$\gamma_{X(S)}$	Activity coefficient of X in a seawater medium
β	Geometric factor
β''	Function of the Pitzer ionic interaction parameters
ϕ and ϕ''	Second virtual coefficient
Ψ and Ψ''	Third virtual mixing parameter in Pitzer equation

5.8   References

1.   Wilf, M., Development of new technologies for the reduction of fouling and improvement of performance in Seawater RO. 2002.
2.   Hannachi, A., et al. A new index for scaling assessment. in IDA World Congress-Maspalomas. 2007. Gran Canaria.
3.   Langelier, W., Chemical equilibria in water treatment. J. AWWA 1946. 38(2): p. 169.
4.   Stiff, H.A. and L.E. Davis, A method for predicting the tendency of oil field waters to deposit calcium carbonate. Petroleum transactions, 1952. 195: p. 213-216.

5.   Butt, F.H., F. Rahman, and U. Baduruthamal, Characterization of foulants by autopsy of RO desalination membranes. Desalination, 1997. 114: p. 51-64.

6.   Butler, J., Carbon Dioxide equilibria and their applications. second ed. 1982, California: Addison-Wesley.

7.   Sohnel, O. and J. Garside, Precipitation basis principals and industrial applications. 1992, Oxford: Butterworth-Heinemann.

8.   ASTM, Calculation and Adjustment of the Stiff and Davis Stability Index for Reverse Osmosis. 2001, ASTM International: West Conshohocken, PA, United States.

9.   Neev, D. and K.O. Emery, The Dead Sea; depositional processes and environments of evaporites. Makhon ha-ge'ologi (Israeal). 1967, Jerusalem: National government publication.

10.   Amit, O. and Y.K. Bentor, pH-dilution curves of saline waters Chemical Geology 1971. 7(4): p. 307-313.

11.   Hammer, U.T., Parker, R.C., Limnology of a perturbed highly saline Canadian lake. Archiv fur Hydrobiologie, 1984. 102(1): p. 31-42.

12.   Hammer, U.T., Saline lake ecosystems of the world. Monographiae biologicae. Vol. 59. 1986, Dordrecht: Dr. W.Junk Publishers.

13.   Hansson, I., A new set of acidity constants for carbonic acid and boric acid in seawater. Deep sea research, 1973. 20: p. 461-491.

14.   Mehrbach, C., et al., Measurement of the apparent dissociation constants of the carbonic acid in seawater at atmospheric pressure. Limnol. Oceanogr., 1973. 18: p. 897-907.

15.   Goyet, C. and A. Poisson, New determination of carbonic acid dissociation constants in seawater as a function of temperature and salinity. Deep sea research, 1989. 36(11): p. 1635-1654.

16.   Roy, R.N., et al., The dissociation constants of carbonic acid in seawater at salinities 5 to 45 and temperatures 0 to 45°C. Marine Chemistry, 1993. 44: p. 249-267.

17.   Lee, K. and F.J. Millero, Thermodynamic studies of the carbonate system in seawater. Deep-Sea Research I 1995. 42(11/12): p. 2035-2061.

18.   Lueker, T., A. Dickson, and D. Keeling, Ocean pCO_2 calculated from dissolved inorganic carbon, alkalinity, and equations for K_1 and K_2: validation based on laboratory measurements of CO_2 in gas and seawater at equilibrium. Marine Chemistry 2000. 70(105-119).

19.   Prieto, F.J.M. and F.J. Millero, The values of pK_1+pK_2 for the dissociation of carbonic acid in seawater. Geochemica et Cosmochemica Acta, 2002. 66(14): p. 2529-2540.

20.   Millero, F.J., et al., Dissociation constants of carbonic acid in seawater as a function of salinity and temperature. Marine Chemistry, 2006. 100: p. 80-94.

21.   Butler, J., Ionic Equilibrium: Solubility and pH calculations. 1998, New York: John Wiley & Sons.

22.   Thurmond, V. and F.J. Millero, Ionization of Carbonic Acid in Sodium Chloride Solutions at $25°C$. Journal of Solution Chemistry, 1982. 11(7): p. 447-456.

23.   Millero, F., et al., The dissociation of carbonic acid in NaCl solutions as a function of concentration and temperature. Geochimica et Cosmochimica Acta 2007. 71 p. 46-55.

24.   Harned, H.S. and F.T. Bonne, The First Ionization of Carbonic Acid in Aqueous Solutions of Sodium Chloride. J. Am. Chem. Soc., 1945. 68: p. 1026-1031.

25.   Harned, H.S. and R.D. Jr., The Ionization Constant of Carbonic Acid in Water and the Solubility of Carbon Dioxide in Water and Aqueous Salt Solutions from 0 to 50°C. J. Am. Chem. Soc., 1943. 65(10): p. 2030-2037.

26.   Harned, H.S. and S.R.S. Jr., The Ionization Constant of HCO_3^- from 0 to 50°C. J. Am. Chem. Soc., 1941. 63(6).

27.   Huff, G.F., Use of simulated evaporation to assess the potential for scale formation during reverse osmosis desalination. Desalination, 2004. 160: p. 285-292.

28.   Copeland, B.J., Environmental characteristics of hypersaline lagoons. Contrib. Mar. Sci,, 1967. 12: p. 207-218.

Chapter 6

The role of inorganic ions in the calcium carbonate scaling of seawater reverse osmosis systems

Based on parts of:
Waly, T.; Kennedy, M.K.; Witkamp, G.J.; Amy, G. and Schippers, J.C. The role of inorganic ions in the calcium carbonate scaling of seawater reverse osmosis systems. Desalination, submitted
Waly, T.; Saleh, S.; Kennedy, M.K.; Witkamp, G.J.; Amy, G. and Schippers, J.C. Reducing the calcite scaling risk in SWRO: role of Mg^{2+} & SO_4^{2-}. In proceeding of the IDA world congress. 2009. Dubai, UAE

6.1   Abstract

In supersaturated solutions the period preceding the start of 'measurable' crystallization is normally referred to as the 'induction time'. This research project aimed to investigate the induction times of $CaCO_3$ in the presence of Mg^{2+} and SO_4^{2-} for synthetic seawater, with a focus on the concentrate bulk solution. Mg^{2+} and SO_4^{2-} ions, which are naturally present in seawater, have been reported to disturb the crystal growth of $CaCO_3$ crystals by exchanging Mg^{2+} and SO_4^{2-} with Ca^{2+} and CO_3^{2-}, respectively, hindering their growth and altering the crystal shape. Therefore, the effect of Mg^{2+} and SO_4^{2-} on the induction time of calcium carbonate precipitation in SWRO systems was investigated. As induction time depends on the detection technique, the induction time in this research is defined as a drop of 0.03 pH units which in this research corresponds to 0.1-0.27 mg/L of precipitated calcium carbonate. A highly sensitive online pH meter was employed to measure pH change due to $CaCO_3$ precipitation. Four synthetic water solutions were prepared to represent $CaCO_3$ supersaturated solutions: without Mg^{2+} and SO_4^{2-}; with SO_4^{-2} only; with Mg^{2+} only; and with both Mg^{2+} and SO_4^{2-}. The prepared synthetic solutions have the same ionic strength values found in the Gulf of Oman SWRO concentrates at 30% and 50% recovery. The results showed a significant increase in the induction time by 333%, 457% and 667% for a recovery of 30%, when adding SO_4^{2-} only, Mg^{2+} only, or both Mg^{2+} and SO_4^{2-}, respectively, to synthetic SWRO concentrate compared to that obtained in the absence of Mg^{2+} and SO_4^{2-} at an initial pH of 8.3. Similar results were obtained for a 50% recovery as the induction time increased by 1140%, 2820%, and 3880% in the presence of SO_4^{2-}, Mg^{2+}, and Mg^{2+} and SO_4^{2-}, respectively, at an initial pH of 8.3. The increase in the induction time in the presence of SO_4^{2-} was more than likely to be due to nucleation and growth inhibition while the presence of Mg^{2+} affected the nucleation and growth through both complexation and inhibition. After a 5-month solution stabilization period, ESEM and XRD analyse showed aragonite in solutions containing Mg^{2+}. On the contrary, calcite was the final crystal phase formed in solutions with no Mg^{2+}. This suggest that magnesium may play an important role in inhibiting the formation of calcite.

Keywords: Scaling, Induction time, $CaCO_3$ polymorphs, SWRO, Mg^{2+}, SO_4^{2-}

6.2   Background

Introduction

Calcium carbonate scaling is one of the major concerns in seawater reverse osmosis (SWRO) processes, where saturation indices, e.g., Stiff and Davis Stability Index, Calcium Carbonate Precipitation Potential CCPP and Saturation Index (SI), are widely used for the prediction of $CaCO_3$ scaling in RO systems [1].

Although these commonly used indices are used to estimate solution saturation with respect to $CaCO_3$, they cannot predict the supersaturated solution's stability after its saturation limits are exceeded. This is referred to in the literature as 'induction time' [2]. In this research the aim is to determine the effect of Mg^{2+} and SO_4^{2-} on induction time in high ionic strength medium. It has been reported that constituents of natural seawater such as sulphate, magnesium and natural organic matter (NOM) play a role in calcium carbonate scaling [3-5]. The presence of dissolved ions such as sulphate and magnesium are hypothesized to alter the kinetics and the driving force of precipitation (based on activity) of calcium carbonate [6-8]. Therefore, in this research an attempt was made to incorporate these effects on the indices measuring $CaCO_3$ saturation.

Calcium carbonate phases

The initial solid phase formed by precipitation may not be the thermodynamically stable phase for the reaction conditions. If this is the situation, then over a period of time, the (crystal) structure of the precipitate may change to that of the more stable phase. This change will be accompanied by additional precipitation and consequently a reduced solution concentration because the more stable phase has a lower solubility than the initially formed phase [9-13]. The crystal aging normally includes changes in crystal structure over time in addition to ripening where, the crystal size of the precipitate increases [14]. Crystal growth reported for individual solids may vary widely due to the effects of particle size and aging, and factors such as complex formation, adsorption of impurities at the crystal interface and formation of solid mixtures [15-16].

Polymorphs are minerals with the same chemical composition but a different arrangement of the ions in the crystal lattice (Figure 1). The compound $CaCO_3$ exists in six forms (Table 1) that are divided into two classes: well-known anhydrous crystalline polymorphic forms (calcite, aragonite and vaterite) and, hydrated crystalline forms (calcium carbonate monohydrate (monohydrocalcite), calcium carbonate hexahydrate (ikaite) and amorphous calcium carbonate (ACC) [17]. Calcite is the thermodynamically most stable form of $CaCO_3$ at ambient temperature [18]. Aragonite is thermodynamically less stable than calcite but also often occurs in biological and geological samples. It has an orthorhombic crystal structure with the same alternating structure as calcite [19-20]. At temperatures greater than 50°C aragonite is the most abundant species. In the range of 20-50°C vaterite become the most abundant but it is unstable and it transforms in the lower temperature range (e.g. <30 °C) to calcite and to aragonite in the higher temperature range (e.g. >30 °C) [5, 21]. Vaterite is rare in nature and has a hexagonal crystal structure with a diffrent orientation of the CO_3^{2-} compared to that of calcite and aragonite [19].

Table 1: Solubility of different calcium carbonate forms [22]

Species	pK$_{sp}$ (T in °K)	Range of validity
Calcite	171.9065 + 0.077993.T -2839.319/T- 71.595.log(T)	273.1<T<263.1°K
Aragonite	171.9773 + 0.077993.T -2903.293/T-71.595.log(T)	273.1<T<263.1°K
Vaterite	172.1295 + 0.077996.T -3074.688/T- 71.595.log(T)	273.1<T<263.1°K
Amorphous	6.1987 + 0.0053369.(T-273.15) + 0.0001096.(T-273.15)^2	273.1<T<228.1°K
Monohydrated	7.05 + 0.000159. .(T-273.15)	273.1<T<223.1°K
Hexahydrated	2011.1/T - 0.1598	283.1<T<298.1°K

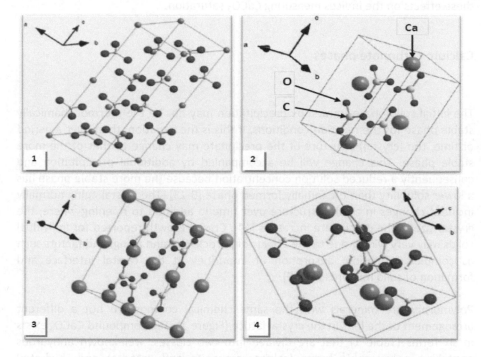

Figure 1: Schematic representation of the crystal structure of calcium carbonate polymorphs:(1) calcite, (2) aragonite, (3) vaterite, (4) calcium carbonate monohydrate [23]

The other three hydrated calcium carbonate polymorphs are unstable. Calcium carbonate monohydrate crystals are mostly spherical in shape with a diameter close to 100 μm and a density of 2.38 g cm^{-3} [20, 24]. Calcium carbonate hexahydrate (ikaite) has a monoclinic structure with Ca^{2+} bound more closely to the six H$_2$O molecules than to the CO$_3$$^{2-}$ ion, and has a density of 1.8 g cm^{-3} [20, 24]. The presence of phosphate suppresses the growth of anhydrous CaCO$_3$ crystals enabling the growth of ikaite [20]. Therefore, experimental procedures for ikaite synthesis usually include the presence of a substance inhibiting the formation of anhydrated phases such as magnesium or polyphosphate [20]. The last hydrated form is known as amorphous calcium carbonate (ACC) which is highly unstable and rapidly transforms into a crystalline phase, and is difficult to characterize due to its short life. Certain organic and inorganic additives have been used to inhibit the transformation of ACC. Magnesium and organic matter show the ability to retard the transformation of ACC into a crystalline phase through incorporation within amorphous calcium carbonate [7, 9-10].

The role of inorganic ions

Although calcite is considered to be the thermodynamically the most stable form of calcium carbonate, the presence of foreign ions sometimes changes this rule [2]. Aragonite was reported to be prevented from transforming into the more stable calcite when certain impurities such as Mg^{2+}, Ni^{2+}, Co^{2+}, Fe^{3+}, Zn^{2+} and Cu^{2+} are present, all of which encourage aragonite formation over calcite, whereas the presence of Mn^{2+}, Cd^{2+}, Ca^{2+}, Sr^{2+}, Pb^{2+} and Ba^{2+} favour calcite formation. Other ions, e.g., SO_4^{2-} and Cl^- were reported to increase the solubility of $CaCO_3$ while, PO_4^{-3} was reported to favour the formation of hexahydrated calcium carbonate over other forms [25-34]. This effect was also reported for the interaction between $CaCO_3$ and toxic elements such as cadmium (Cd) [35] and radionuclides [36].

Mg^{2+} has the strongest influence on $CaCO_3$ precipitation, favouring the formation of aragonite or vaterite over calcite, and inhibiting the phase transformation of vaterite and aragonite to calcite [26, 37]. It was notable that the addition of the magnesium ion at a Mg^{2+}: Ca^{2+} level of 5:1 gave the highest inhibition for $CaCO_3$ crystal formation over other ratios [38]. Studies of low ionic strength waters claimed that even much smaller Mg^{2+}: Ca^{2+} ratios of 0.8:1 tripled the induction time of $CaCO_3$ compared to a lower ratio of 0.4:1 [26, 37]. On the contrary, Loste et al, 2003 [7] demonstrated that amorphous calcium carbonate (ACC) was the first phase formed in the presence of magnesium. The presence of magnesium stabilized and retarded ACC in its transformation to calcite [39]. Previously, this effect has only been observed for ACC stabilized with organic additives. These results have relevance for the formation mechanism of biological magnesium calcites, and suggest that the stabilization of ACC by magnesium may provide organisms with a mechanism for controlling crystal morphologies [7].

In unseeded experiments of calcium carbonate in artificial seawater, vaterite and aragonite polymorphs are known to extensively occur during calcium carbonate precipitation from aqueous solutions with Mg^{2+} concentrations similar to that in seawater environments [15, 40-45]. Without Mg^{2+} most likely calcite is found. Furthermore, vaterite was reported as the first precipitate at Mg^{2+}:Ca^{2+} ratios higher than 3:1 [15, 44-46].

Research has demonstrated that Mg^{2+} and SO_4^{2-} cause a decrease in the $CaCO_3$ growth rate by a factor of 25 times compared with that experienced in their absence at the same Ca^{2+} and CO_3^{2-} activities. In the presence of Mg^{2+}, the formation of dolomite is still a matter of debate, but it was agreed that Mg^{2+} replaced nearly 10-30% of the Ca^{2+} in the $CaCO_3$ crystal lattice [46-48]. Whether this effect is due to complexation or a decrease in activity is not known.

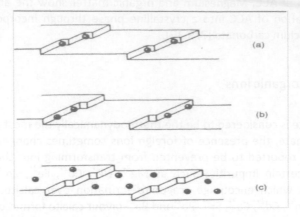

Figure 2: Sites for trace element-adsorption on growing flat crystal faces [49]

Mg^{2+} was also shown to have an effect on the growth of calcium phosphates in seawater and Mg^{2+} bearing solutions. This inhibitory effect was explained by blocking of the surface growth sites of calcium phosphate crystals by adsorbed Mg^{2+} ions as shown in Figure 2 [50]. This phenomenon was reported for other calcium containing crystals such as brushite ($CaHPO_4.2H_2O$) where the Mg^{2+} presence inhibited its transformation to the more stable forms such as octacalcium phosphate and apatite [51]. On the other hand, Ca^{2+} was reported to have an inhibition effect and prolong the induction time of magnesium-containing compound such as $MgNH_4PO_4 . 6H_2O$ [52].

Unlike the effect of cations on $CaCO_3$ precipitation, less research has been dedicated to study of the effect of anions (e.g. sulphate and, silica) [53]. They have been reported to influence the precipitation even very low concentrations [54]. The presence of these former ions suppressed the crystallization of $CaCO_3$, as well as affecting the morphology of the formed $CaCO_3$ crystals [55]. Chong and Sheikholeslami, 2001 [6] illustrated that at ratios of $SO_4^{2-}:CO_3^{2-}$ of nearly 0.35:1, the heat of reaction of $CaCO_3$ is affected in the presence of SO_4^{2-} compared to solutions without SO_4^{2-}. The $CaCO_3$ scale formed was less dense when compared to that in the absence of SO_4^{2-}. They illustrated as well that kinetic equations for single salt precipitation are different than in a mixed salt environment.

Although pilot test studies are rare, an interesting field pilot test on coal mine water confirmed the positive effect of Mg^{2+} and SO_4^{2-} on $CaCO_3$ solubility. The increase in solubility was proposed to be by a factor 3 without adding any acids or antiscalants [56]

Solubility of salts

When an ionic compound is added to water, it will usually (partially) dissolve in the solution as its ions. If the activity of an ionic compound added to a volume is in excess, equilibrium is reached when the number of ions entering the solution from the solid compound is equal to the number of ions leaving the solution to the solid compound according to the following reaction [57]:

$$CaCO_3 \rightleftharpoons Ca^{2+} + CO_3^{2-} \qquad\qquad 1.$$

The first requirement for precipitation or scale formation is supersaturation of the solution with respect to the scaling salt. If the solubility is exceeded based on the maximum activity of salt soluble in a solution (at a given temperature) determined by the equilibrium thermodynamic solubility product K_{sp}, precipitation may occur [2, 35, 58-60]. K_{sp} varies with temperature wherein alkaline scale solubility decreases as the temperature increases [61]. It is also noted that scaling may be enhanced by surface roughness, hydraulic conditions, as well as the surface charge [62].

The K_{sp} (thermodynamic solubility product) can be determined using both laboratory experiments employing the seeded growth technique [58, 60] and incorporating calculated activity coefficients (equations 2 and 3) [4, 58, 63-64].

$$K_{sp} = [Ca^{2+}][CO_3^{2-}]\gamma_+\gamma_- \qquad\qquad 2.$$

$$pK_{sp} = pK_{so} + log\,\gamma_+ + log\,\gamma_- \qquad\qquad 3.$$

Where K_{sp} is the thermodynamic solubility product, K_{so} is the solubility in pure water , $\gamma+$ is the cation activity coefficient and $\gamma-$ is the anion activity coefficient

Pitzer [65-68] have taken into account the effect of ion pairing on the calculations of the activity coefficients for high ionic strength waters, e.g., 6 molar. Using this approach in solubility product calculations enables researchers to model real scalant behaviour in real saline waters [58].

Common methods used to determine the precipitation potential of CaCO$_3$:

In seawater CaCO$_3$ saturation is calculated by the use of different indices such as Stiff & Davis Stability Index (S&DSI) [69]; Saturation Index (SI) and Saturation Ratio (S$_a$). These indices were developed to predict the tendency of sparingly soluble salts in general, except for S&DSI, which is developed specifically for CaCO$_3$.

Saturation indices (SI and S$_a$)

The saturation index (SI) (Equation 4) predicts the scaling potential of sparingly soluble salts taking into consideration the interaction between ions. The index incorporates the activity coefficients in its activity calculations to take into account the effect of ionic complexation due to high salinity. For SWRO concentrates it is preferable to use Pitzer to calculate the activity coefficients rather than Davis or extended Debye Huckel activity coefficients [2, 22, 55, 58, 65, 67, 70].

$$SI = Log\left(\frac{[Ca^{2+}]\gamma_{ca^{2+}}[CO_3^{2-}]\gamma_{CO_3^{2-}}}{K_{sp}}\right)$$ 4.

Using a similar concept, the saturation ratio (S$_a$) can be written as follows (Equation 5)

$$S_a = \frac{[Ca^{2+}]\gamma_{ca^{2+}}[CO_3^{2-}]\gamma_{CO_3^{2-}}}{K_{sp}}$$ 5.

Mechanism of nucleation and growth

Theoretically, once supersaturation is achieved, precipitation is possible however, supersaturation is not the only factor that is involved in scale formation as the kinetics of scale formation are also important. Nucleation starts by the formation of clusters which become nuclei under a homogenous or heterogeneous nucleation mechanism [71]. Homogeneous nucleation is where the formation of the solid phase is not influenced by the presence of any solid phase, and heterogeneous nucleation is where the formation of new solid phase particles is catalyzed by the presence of a solid particles [2]. These nuclei grow further to the crystalline phase while new nuclei are still forming. In other cases, it is initiated by the presence of the solid phase already formed from the crystallizing material itself (secondary nucleation) [2]. The formation of stable crystals sometimes needs an aging step in which coagulation and sedimentation contribute to the formation of the final product [16, 72].

Literature values for time intervals (t$_n$) between nucleation events for homogenous nucleation were reported to be from 10^{-2} to as low as 10^{-7} seconds [2, 37, 49]. Meanwhile, the presence of magnesium was reported to increase these values about 100 times compared to control solutions [37]. In general, the presence of foreign particles affects both the time needed for nucleation (t$_n$) and that needed for subsequent growth (t$_g$) to a detectable size. The differentiation between these two times depends largely on the technique used, but it may also be useful to combine induction time measurements with that of growth rate in a seeded growth experiment. [73-74].

Induction time

In practice the period of metastability preceding the detection of the precipitation process is commonly indicated as the induction time (t_{ind}) [2, 37, 58, 75]. As it is very difficult to determine the formation of the first nuclei experimentally, consequently, a part of the experimentally-measured induction time may also include growth to a detectable size [2].

In case of a steady state nucleation rate (J_s) and nucleation time (t_n)>> growth time (t_g), the induction time can be expressed as in equations 6-11 [2].

$$t_{ind} \propto J_S^{-1}$$ 6.

Therefore, the induction time can be expressed as follows (Equation 7)

$$t_{ind} = \frac{1}{\Omega} exp\left(\frac{\beta\vartheta^2\gamma_s^3 f(\theta)}{k_b T\emptyset^2}\right)$$ 7.

Where β is the geometric factor for nucleation, ϑ Molecular volume, γ_s is the surface energy, Ω Pre-exponential factor, T is the temperature in Kelvin and k_b is the Boltzmann constant

In equation 7, The value of $f(\theta)$ determines the nucleation mechanism. Homogeneous nucleation is recognized by $f(\theta)$ =1 while for heterogeneous nucleation $f(\theta)$ <1. Values as low as 0.01 were reported for $f(\theta)$

Equation 7 can be written in a simplified way as follows:

$$logt_{ind} = \frac{B}{(T)^3.log^2S_a} - A$$ 8.

Where

$$B = \left[\frac{\beta\vartheta^2\gamma_s^3 f(\theta)}{v^2(2.3k_b)^3}\right]$$ 9.

And

$$A = log\Omega$$ 10.

$$\emptyset = vk_b TLnS_a$$ 11.

Where υ is the number of ions into which a molecule dissociates and S_a is the supersaturation ratio.

If induction time is dominated by growth and $t_g >> t_n$ then the expression for induction time which in that case represent the growth time, can be written as according to equations 12, which represent the screw-dislocation growth mononuclear growth [2, 73-74].

$$t_{ind} = \frac{d^3}{6Dr^*} exp\left(\frac{\beta'\vartheta^{4/3}\gamma_s^2}{k_bT\emptyset}\right)$$ 12.

Where D is the diffusion coefficient in the solution, r* is the critical nuclei radius, d Interplaner distance in solid phase and β' is the geometric factor for growth.

The relations represented assume that nucleation is followed by growth. However, in reality growth and nucleation are occurring simultaneously. In this case, the supersaturation of the solution hardly decreases at the designated induction time and therefore, the force for nucleation will remain constant. In such a case, nuclei keep forming while those formed keep growing. Based on this assumption, the induction time can be rewritten as a relation between growth rate (G_g) and nucleation rate (J) [76-78] as in equation 13.

$$t_{ind} \propto (G_g^3.J)^{\frac{-1}{4}}$$ 13.

Meanwhile, The steady state growth rate and the steady state nucleation rate are correlated to the time for growth and time of the formation of the first nuclei as in equations 14-15 [2]

$$t_g \propto G_g^{-1}$$ 14.

And

$$t_n \propto J^{-1}$$ 15.

Therefore, equation 16 can be written by substituting of equation 14 and 15 in equation 13.

$$t_{ind} \propto (t_g^3 . t_n)^{\frac{1}{4}}$$ 16.

But since nucleation time and growth time are correlated to $\emptyset$ which in turn is correlated to the saturation ratio as shown in equations 8 and 12 equation 16 can be written as a function of the saturation index as shown in equation 17

$$log t_{ind} = k - n log(S_a) = k - nSI$$ 17.

Where k is a constant with no physical meaning and n is the order.

Based on the previous equation (equation 17), if the induction time includes both nucleation and growth (they happens simultaneously), and the experiments covers a limited range of supersaturation (no change in the nucleation mechanism is expected and therefore in the surface energy), a linear fit between the logarithm of the induction time and the logarithm of saturation ratio can be expected [2, 50, 79-80].

6.3   Materials

pH meter

The induction time measurements using pH were performed with a highly sensitive pH meter (Eutech pH 6000) which has an accuracy of 0.001 pH units. The pH meter is connected online for continuous measurement of pH over time. The pH probe was fitted in the top of the experimental glass reactor, which has a volume of 3 litres (Applikon). The pH measurements were performed online using the manufacturer's software or offline by using the instrument memory and the measuring interval can be adjusted to as low as every 30 seconds.

Reactors

The pH probe was fitted in the top of an air-tight double-jacketed glass reactor with a volume of 3 litres (Applikon), and equipped with a double-paddled shaft mechanical stirrer. The mixing rate was varied from 0 to 1,200 rpm using an electronic controller (Applikon) linked to the mixing motor (Figure 3). The reactor can be filled either manually or mechanically using a diaphragm pump with an average filling speed of 4 L/min.

Figure 3: The reactor's assembly and setup

After each experiment cleaning employed 0.2 molar HNO_3 for 30 minutes with a flow of 0.15 L/min to dissolve any formed crystals. The reactor was then flushed with demineralised water for 15 minutes with a flow rate of 3 L/min before the next experiment.

Synthetic seawater concentrate preparation

The synthetic seawater concentrate used in the experiments was prepared in stages using ultra-pure water. The ultra-pure water system made use of tap water as the raw water source where it passed through a series of treatment steps to decrease the organic and inorganic particle content in the feed water. The product water had a conductivity and total organic carbon (TOC) of 0.8 µS/cm and 3 µg/L, respectively. The TOC was measured using the TOC analyzer with a detection limit of 0.5 µg/L.

Table 2: Feed and concentrate seawater composition at 30% and 50% recovery for the desalination plant in the Gulf of Oman

Parameter	Unit	Feed	R=30%	R=50%
Ammonium	mgN/L	0.03	0.04	0.06
Calcium	mg/L	474	677	948
Sodium	mg/L	12,245	17,492	24,489
Magnesium	mg/L	1,356	1,937	2,711
Potassium	mg/L	434	620	868
Phosphate	mgP/l	0.08	0.35	0.5
Silicate	mg/L	0.13	0.19	0.26
Chloride	mg/L	21,535	30,764	43,070
Sulphate	mg/L	2,772	3,960	5,544
Bicarbonate	mg/L	146	209	293
TDS	ppm	39,017	55,739	78,034
Ionic strength	**Mole/L**	**0.78**	**1.12**	**1.61**

Table 3: Chemical concentrations for preparing the synthetic solutions of 30% recovery

Chemical salts	$SO_4^{2-}+Mg^{2+}$	Only SO_4^{2-}	Only Mg^{2+}	No SO_4^{2-} or Mg^{2+}
	mg/L	mg/L	mg/L	mg/L
$NaHCO_3$	288	288	288	288
$CaCl_2.2H_2O$	2,483	2,483	2,483	2,483
$MgCl_2.6H_2O$	16,191	0.0	16,191	0.0
Na_2SO_4	5,856	5,856	0.0	0.0
NaCl	36,277	50,247	48,324	62,294
Ionic strength (mole/L)	1.12	1.12	1.12	1.12

The chemical dosages were calculated for preparing the synthetic seawater based on the concentrate composition for a SWRO plant using seawater from the Gulf of Oman (Table 2). The concentrate preparation was done to simulate the concentrate of the SWRO at a working recovery of 30% and 50% (Tables 3 and 4). For each recovery, four batch solutions were prepared using high grade analytical salts (Table 5) to simulate the effect of the presence of Mg^{2+} and SO_4^{2-} on the nucleation mechanism of $CaCO_3$. The ionic strength of the four batches at each recovery was kept constant at 1.12 and 1.61 mole/L.

The four different batch solutions were prepared by either excluding the SO_4^{2-} (the second solution in Tables 3 and 4) or by excluding Mg^{2+} (the third solution in Tables 3 and 4) or both (the fourth solution in Tables 3 and 4). The first solution was prepared to determine the overall effect of both ions on the $CaCO_3$ induction time.

Table 4: Chemical concentrations for preparing the synthetic solutions of 50% recovery

Chemical Salts	$SO_4^{2-}+Mg^{2+}$	Only SO_4^{2-}	Only Mg^{2+}	No SO_4^{2-} or Mg^{2+}
	mg/L	mg/L	mg/L	mg/L
NaHCO$_3$	403	403	403	403
CaCl$_2$.2H$_2$O	3,477	3,477	3,477	3,477
MgCl$_2$.6H$_2$O	22,668	0.0	22,668	0.0
Na$_2$SO$_4$	8,198	8,198	0.0	0.0
NaCl	53,242	72,800	70,108	89,665
Ionic strength (mole/L)	1.61	1.61	1.61	1.61

The synthetic seawater concentrate used in the experiments was prepared in stages. Firstly, a $NaHCO_3+Na_2SO_4$ solution was prepared by dissolving $NaHCO_3$ and Na_2SO_4 salts (Table 5) in ultra-pure water. Secondly, a $CaCl_2.2H_2O$ +$MgCl_2.6H_2O$ + NaCl solution was prepared by dissolving three of the salts in sequence.

Table 5: Salt reagents used in the experimental synthetic seawater concentrate preparation

Reagent name	Reagent composition	Form	Grade
Sodium bicarbonate	NaHCO$_3$	Salt	98.0%
Calcium chloride dihydrate	CaCl$_2$.2H$_2$O	Salt	99.0%
Magnesium Chloride hexahydrate	MgCl$_2$.6H$_2$O	Salt	98.5%
Sodium Sulphate	Na$_2$SO$_4$	Salt	99.0%
Potassium Chloride	KCl	Salt	99.5%
Sodium Chloride	NaCl	Salt	99.5%

To ensure a complete dissolution of the reagents, in the preparation step the solutions were dissolved in batches of 1 L. The salt was added to the ultra-pure water and the flask was then closed and shaken manually for 2 minutes after which

2 hours of solution mixing took place on a magnetic stirrer. Mixing was performed at an average speed of 400 rpm and at a room temperature of 20 °C.

6.4 Methods

Induction time measurements

The induction time experiments were initiated by adding the $NaHCO_3$ + Na_2SO_4 solution into the reactor followed by the NaOH solution for pH correction (if needed). Finally the $CaCl_2.2H_2O$ + NaCl + $MgCl_2.6H_2O$ solution was added at a rate of 0.2 L/min while maintaining a mixing speed of 150 rpm to ensure proper mixing and to minimize the chance of the formation of local supersaturation zones. The addition was performed through fine nozzles located 3 cm from the reactor's base to ensure proper distribution of the solution when added. The two reacting solutions were added on a 1:1 volume basis.

The pH drop was monitored over a period of 2,000 minutes, and the induction time was defined (in this research) as a pH drop of at least 0.03 pH units. The amount of $CaCO_3$ precipitate after this pH drop is dependent on the initial solution saturation and the corresponding alkalinity. Therefore, in the range of saturation covered in this research, the indicated drop in pH is equivalent to 0.1-0.25 mg/L of $CaCO_3$.

Determination of co-precipitation

In order to make sure that brucite ($Mg(OH)_2$), dolomite ($MgCa(CO_3)_2$) and magnesite ($MgCO_3$) did not precipitate with $CaCO_3$ and therefore influence the induction time measurements, samples were taken from the crystals formed on the reactor's walls for analysis. The first step was to determine the ratio between Ca^{2+} and Mg^{2+} in the precipitating crystals using inductively-coupled plasma (ICP). The procedure involved scraping samples from the reactor wall, which were rinsed 5 times with ultra-pure water and then re-dissolved using 0.1 M HCl solution. The samples were measured using ICP to determine the ratio between the Mg^{2+} and the Ca^{2+} in the precipitated crystals.

Determination of the final crystal phase of calcium carbonate

Different samples for the 30% and 50% recovery solutions were collected from the reactors at the end of the experiments in order to examine the calcium carbonate crystals formed at the end of the experiment (assumed to be the final phase). At the end of the experiments (48 hours from starting), the supernatant was removed, and the particles were collected then and washed by ultra-pure water then stored in ultra-pure water at 20°C for 6 months to ensure full crystal aging and ripening before being investigated by the environmental scanning electron microscope (ESEM) and x-ray diffraction (XRD).

An environmental scanning microscope (ESEM) (Figure 4) is a scanning electron microscope (SEM) with an extra feature that allows a gaseous environment in the specimen chamber (scanning electron microscopes normally operate in a vacuum). The presence of gas around a specimen creates a new possibility unique to ESEM where hydrated specimens can be examined since any pressure greater than 609 Pa allows water to be maintained in its liquid phase for temperatures above 0 °C, in contrast to the SEM where specimens are desiccated by the vacuum conditions.

Figure 4: Environmental Scanning Electron Microscope

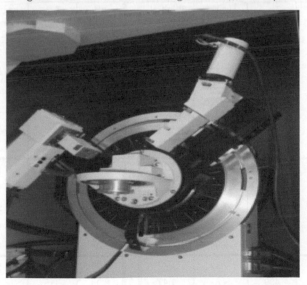

Figure 5: X-ray diffraction instrument

X-ray diffracion (XRD) (Figure 5) was employed to identify the crystal phase formed
and to confirm the ESEM results. The latter depends on irradiating a dry powdered
sample of a crystalline material with X-rays of a known wavelength. The diffracted
X-rays are detected as a function of their diffraction angle, which is related to the
position of the lattice planes in the crystal. This results in a pattern of peaks versus
diffraction angles (diffraction pattern), in which positions and intensities are typical
for a crystalline compound. Identification of the crystalline compound(s) takes
place using an extensive database of diffraction patterns for different crystalline
phases, or by full-profile-fitting procedures.

6.5   Results and discussion

pH versus induction time (T_{ind})

The results represented in Tables 6 and 7 suggest that, in general, the induction
time decreases as the experimental initial solution pH increases. Figures 6 and 7
confirmed this relationship for 30% and 50% recovery and a logarithmic
relationship between the initial pH and the induction time is proposed with a
regression coefficient (R^2) ranging from 0.95-99. It is worth mentioning that
solutions containing Mg^{2+} or both Mg^{2+} and SO_4^{2-} showed no signs of crystallization
over the whole experimental time span (2,000 min) when initial pH values were
less than 8.15 and 8.0 at 30% and 50% recovery, respectively.

Table 6: Summary of the experimental initial conditions vs. induction time for
supersaturated synthetic solutions of $CaCO_3$ simulating SWRO concentrates of 30% recovery

Exp. Set	pH (initial)	T_{ind} (min)	Log IAP (initial)	SI(initial) (calcite)	pH (initial)	T_{ind} (min)	Log IAP (initial)	SI (initial) (calcite)
No Mg^{2+} or SO_4^{2-} (30% R)	7.82	678	-7.64	0.81	8.66	105	-6.92	1.53
	8.00	349	-7.47	0.98	8.79	41	-6.84	1.61
	8.28	184	-7.22	1.23	8.98	26	-6.72	1.73
	8.32	240	-7.19	1.26	9.04	40	-6.69	1.76
	8.50	138	-7.04	1.41	9.10	22	-6.66	1.79
	8.63	100	-6.94	1.51				
Mg^{2+} and SO_4^{2-} (30% R)	8.32	1610	-7.29	1.16	8.74	628	-7.03	1.42
	8.55	1144	-7.14	1.31	8.83	526	-6.99	1.46
	8.55	917	-7.14	1.31	9.04	426	-6.91	1.54
	8.61	950	-7.10	1.35	9.48	162	-6.81	1.64
	8.73	744	-7.04	1.41	9.58	142	-6.80	1.65
Mg^{2+} (30% R)	8.24	1182	-7.32	1.13	8.88	408	-6.95	1.50
	8.68	489	-7.04	1.41	8.93	355	-6.94	1.51
	8.83	492	-6.97	1.48	9.27	152	-6.84	1.61
	8.87	430	-6.96	1.49	9.31	136	-6.83	1.62

	8.30	1248	-7.24	1.21	8.90	70	-6.79	1.66
	8.40	498	-7.16	1.29	9.04	97	-6.72	1.73
SO_4^{2-} (30% R)	8.45	497	-7.12	1.33	9.27	33	-6.61	1.84
	8.51	302	-7.07	1.38	9.29	22	-6.60	1.85
	8.68	257	-6.94	1.51				

Table 7: Summary of the experimental initial conditions vs. induction time results for supersaturated synthetic solutions of CaCO3 simulating SWRO concentrates of 50% recovery

Exp. Set	pH (initial)	T_{ind} (min)	Log IAP (initial)	SI (initial) (calcite)	pH (initial)	T_{ind} (min)	Log IAP (initial)	SI (initial) (calcite)
	7.51	778	-7.58	0.87	8.31	21	-6.87	1.58
	7.65	554	-7.45	1.00	8.34	19	-6.85	1.60
	7.82	168	-7.29	1.16	8.41	20	-6.79	1.66
	7.83	181	-7.28	1.17	8.41	22	-6.79	1.66
	7.91	101	-7.21	1.24	8.43	9	-6.78	1.67
No Mg^{2+} or SO_4^{2-} (50% R)	7.92	114	-7.20	1.25	8.52	11	-6.71	1.74
	8.09	49	-7.05	1.40	8.52	18	-6.71	1.74
	8.09	60	-7.05	1.40	8.53	13	-6.71	1.74
	8.20	55	-6.96	1.49	8.54	10	-6.70	1.75
	8.20	44	-6.96	1.49	8.60	13	-6.66	1.79
	8.23	41	-6.93	1.52	8.62	7	-6.65	1.80
	8.25	28	-6.92	1.53	8.64	4	-6.63	1.82
	8.25	24	-6.92	1.53	8.67	7	-6.61	1.84
	8.27	20	-6.90	1.55				
	8.18	1296	-7.12	1.33	9.15	267	-6.68	1.77
	8.25	927	-7.08	1.37	9.34	262	-6.65	1.80
	8.49	602	-6.93	1.52	9.59	272	-6.63	1.82
Mg^{2+} and SO_4^{2-} (50% R)	8.53	692	-6.91	1.54	9.67	321	-6.63	1.82
	8.75	412	-6.80	1.65	9.78	210	-6.63	1.82
	9.03	202	-6.71	1.74	9.8	376	-6.63	1.82
	9.11	220	-6.69	1.76	9.9	299	-6.63	1.82
	8.19	870	-7.07	1.38	8.86	182	-6.74	1.71
	8.51	460	-6.88	1.57	9.09	280	-6.68	1.77
Mg^{2+} (50% R)	8.53	450	-6.87	1.58	9.37	252	-6.64	1.81
	8.60	302	-6.83	1.62	9.77	272	-6.62	1.83
	8.80	268	-6.76	1.69	9.81	260	-6.62	1.83
	8.04	1278	-7.16	1.29	8.55	76	-6.75	1.70
SO_4^{2-} (50% R)	8.14	728	-7.08	1.37	8.81	38	-6.58	1.87
	8.32	182	-6.93	1.52	9.01	16	-6.48	1.97
	8.50	88	-6.79	1.66				

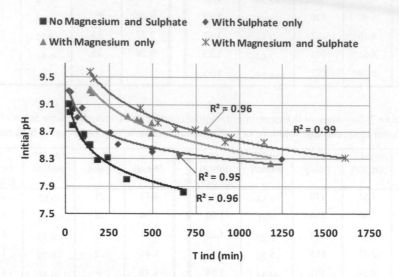

Figure 6: Initial pH versus induction time (min.) for 30% recovery without Mg^{2+} or SO_4^{2}, with SO_4^{2-} alone, with Mg^{2+} alone and with both Mg^{2+} and SO_4^{2-} at 20°C (SI are varying)

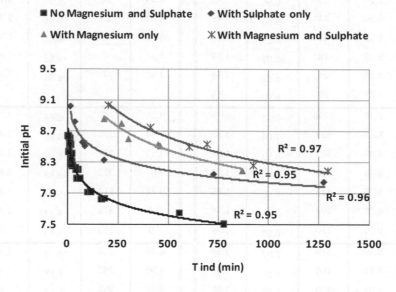

Figure 7: Initial pH versus induction time (min.) for 50% recovery without Mg^{2+} or SO_4^{2}, with SO_4^{2-} alone, with Mg^{2+} alone and with both Mg^{2+} and SO_4^{2-} at 20°C (SI are varying)

The relation between pH and induction time (minutes) is shown in Figures 6 and 7. In the absence of both Mg^{2+} and SO_4^{2-}, the tested solutions show much faster signs of nucleation and growth over the whole range of pH values tested at the same initial concentration (but SI may values may vary). The induction times at 30% and 50% recovery were nearly 8-27 times longer at a pH of 8.3 compared to a pH of 9.0, respectively (Tables 6 and 7). The same pattern was found for solutions containing only SO_4^{2-}, only Mg^{2+} or both Mg^{2+} and SO_4^{2-}.

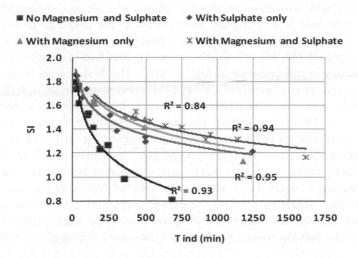

Figure 8: Saturation index versus induction time (min.) for 30% recovery without Mg^{2+} or SO_4^{2-}, with SO_4^{2-} alone, with Mg^{2+} alone and with both Mg^{2+} and SO_4^{2-} at 20°C

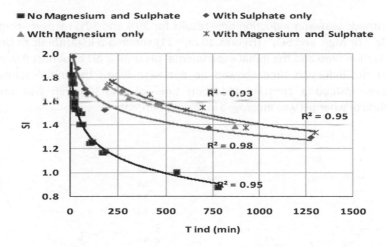

Figure 9: SI versus induction time (min.) for 50% recover without Mg^{2+} or SO_4^{2-}, with SO_4^{2-} alone, with Mg^{2+} alone and with both Mg^{2+} and SO_4^{2-} at 20°C

The relation between induction time (minutes) and the SI presented in Figures 8 and 9 were drawn to show the kinetic effect in the presence of magnesium and sulphate ions. The figures show a logarithmic fit with a correlation coefficient (R^2) of approximately 0.95. The results also show an almost parallel trend for all the lines at the same recovery, except for the control solutions which exhibit a recognizably steeper decline. On the other hand, the relation shows that at a given induction time, the combined presence of Mg^{2+} and SO_4^{2-} shows the strongest inhibiting effect compared to that in the presence of both magnesium, or sulphate alone or in the absence of both magnesium and sulphate. This inhibition effect affected the SI needed to obtain the same induction time. For example for 50% recovery and at induction time of 500 minute, the SI values were 1.41, 1.52 and 1.55 for experiments with SO_4^{2-} alone, with Mg^{+2} alone and with both Mg^{+2} and SO_4^{2-}, respectively. This suggests that the kinetics of a single salt precipitation are different than that of a mixed salt environment in the presence of Mg^{+2} and/or SO_4^{2-} [6] where, their presence significantly affect the reaction rate, thus retarding the crystal growth [44, 46]. It is worth mentioning that two mechanisms may occur simultaneously in the presence of Mg^{+2} or SO_4^{2-}. The first is complexation, which affects the activity coefficients of Ca^{2+} and CO_3^{2-} and results in an overall decrease in available ions for precipitation (corrected by incorporating the activity coefficients in the SI calculations). The second mechanism involved is inhibition, whereby nuclei already formed are inhibited from growing further through growth sites blocking. Consequently, the Mg^{2+} and SO_4^{2-} ions may attach to the freshly formed $CaCO_3$ nuclei, leading to a decrease in the number of sites available for growth causing a decrease in the growth rate of $CaCO_3$ nuclei even at high supersaturation values by blocking the growth steps [49]. Sohnel and Garside [2] went even further and claimed that growth may be suppressed completely especially if foreign ion presence is dominant compared to that of Ca^{2+} and CO_3^2.

Maximum pH:

The synthetic seawater induction time results for 30% and 50% recovery containing only Mg^{2+} or Mg^{2+} and SO_4^{2-} (Figures 10 and 11) showed a logarithmic fit between the induction time and the initial experimental pH until a pH of almost 9.0, beyond this pH the induction time showed no decrease when the initial solution pH increased. Instead a constant induction time of 250-300 min was observed regardless of what pH was employed in the experiment

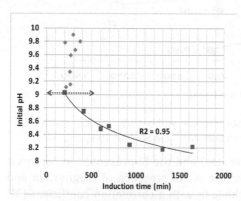

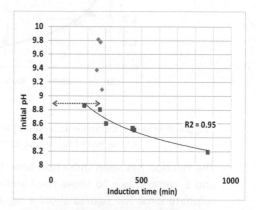

Figure 10: Relation between initial pH and induction time for synthetic seawater at 50% recovery in the presence of Mg^{2+} and SO_4^{2-}

Figure 11: Relation between initial pH and induction time for synthetic seawater at 50% recovery in the presence of only Mg^{2+}

The first assumption is that perhaps this results occurred due to precipitation of magnesium carbonate compounds in the solution e.g. dolomite ($MgCa(CO_3)_2$) or magnesite ($MgCO_3$). Calculations showed that the experimental synthetic solution is highly supersaturated with dolomite and magnesite, while it was just saturated with brucite (Mg(OH)2) at initial pH values higher than 9.35 (See Figure 12 and Table 8).

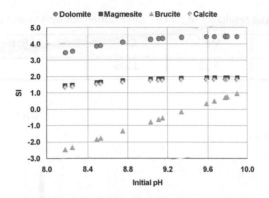

Figure 12: SI vs. pH for synthetic seawater concentrate containing Mg^{2+} and SO_4^{2-} (50% recovery)

Table 8: The initial experimental pH vs. the SI values for calcite; aragonite; brucite, dolomite and magnesite using the Pitzer activity coefficient and calculated using PhreeqC at 20°C

Initial pH	SI				
	Calcite	Aragonite	Brucite	Dolomite	Magnesite
8.18	1.26	1.07	-2.48	3.48	1.43
8.22	1.28	1.09	-2.4	3.51	1.46
8.25	1.3	1.11	-2.34	3.55	1.48
8.49	1.45	1.26	-1.86	3.85	1.63
8.53	1.47	1.28	-1.78	3.89	1.65
8.75	1.58	1.39	-1.34	4.1	1.75
9.03	1.67	1.48	-0.78	4.28	1.87
9.11	1.69	1.5	-0.62	4.32	1.86
9.15	1.7	1.5	-0.54	4.33	1.87
9.34	1.73	1.54	-0.16	4.4	1.9
9.59	1.75	1.56	0.33	4.44	1.92
9.67	1.75	1.56	0.49	4.45	1.93
9.78	1.75	1.56	0.71	4.45	1.93
9.8	1.75	1.56	0.75	4.45	1.93
9.9	1.75	1.56	0.95	4.45	1.93

For the purpose of confirmation of the previous hypothesis, crystal samples were taken from the bottom of the reactors at the end of relevant experiments to measure the Ca^{2+} to Mg^{2+} ratio. Sample analysis was carried out using the Atomic Absorption Spectroscopy (AAS), Inductively Coupled Plasma (ICP) and Environmental Scanning Electron Microscope (ESEM). The average results presented in Table 9 (acquired by ASS and ICP) and Figure 13 (acquired by ESEM) confirm that there was no co-precipitation of Mg^{2+} compounds with $CaCO_3$. The Mg^{2+} content in the crystals was only about 0.5% as much as that of the calcium.

Table 9: Analysis results for Mg^{2+}: Ca^{2+} ratio carried out by AAS and verified by ICP

Element	Mg^{2+}	Ca^{2+}	Ratio (Mg^{2+} : Ca^{2+})
Conc. (mg/l)	5.58	1105	1:198

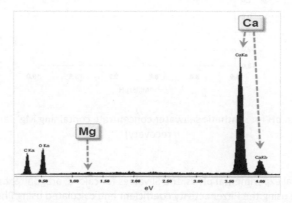

Figure 13: ESEM emission signal showing Ca^{2+}: Mg^{2+} signal ratio

On the other hand, SI calculations presented in Figure 12 and Table 8 showed no increase in the solution saturation when the pH increased beyond 9.34. The start of the SI stability period coincides with the pH at which the induction time cease to decrease as pH increases. These results can be seen in conjunction with Figure 14 which shows the inorganic carbon species in solution with the same ionic strength of the experimental synthetic solution. Figure 14 suggests that above a pH of 9 the increase in CO_3^{2-} in the solution caused by increasing the pH is smaller compared to the lower pH range. At a pH of 9.53 (maximum pH) all the inorganic carbon present in the solution is converted to the CO_3^{2-} species leaving no HCO_3^- to be converted to CO_3^{2-}. Any further increase in the pH did not affect the CO_3^{2-} concentration and resulted in a constant but maximum saturation index value (SI_{max}).

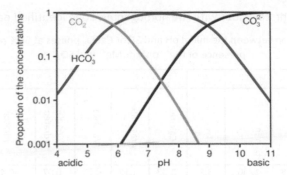

Figure 14 The distribution percentage of the inorganic carbon species in synthetic seawater concentrate of ionic strength equals 1.61 mole/L

These results confirm that the main reason for having a constant induction time at different initial pH values greater than 9.0 is the transformation of all the inorganic carbon in the solution into carbonate and hence resulting in no saturation increase. The results confirm that co-precipitation of Mg^{2+} compounds with $CaCO_3$ is unlikely to happen at SI values less than 0.95, 1.93 and 4.45 for brucite, magnesite and dolomite compounds, respectively.

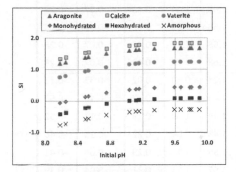

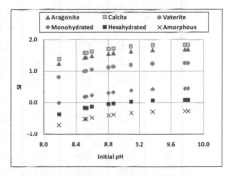

Figure 15: Relation between initial pH and SI for $CaCO_3$ phases at 50% recovery in the presence of Mg^{2+} and SO_4^{2-}

Figure 16: Relation between initial pH and SI for $CaCO_3$ phases at 50% recovery in the presence of only Mg^{2+}

In Chapter 4 the mechanism of nucleation was suggested to be related to the initial solution saturation. Indications showed that homogenous nucleation was dominant when the hexahydrated $CaCO_3$ (ikaite) saturation was exceeded. On the other hand, heterogeneous nucleation suggested to be the dominant mechanism when the solution initial saturation lay between vaterite and monohydrated $CaCO_3$. An intermediate zone with different surface energy values was recognized in between these two zones, where the initial saturation is greater than that of the monohydrated and smaller than that of the hexahydrated $CaCO_3$.

Table 10 and Figures 15 and 16 show that nucleation is expected to be homogenous when the initial pH is greater than 9.03 in the presence of both Mg^{2+} and SO_4^{2-} at 50% recovery (Table 10). A similar value (pH >8.86) was recognized in cases of 50% recovery with only Mg^{2+}. The pH range beyond which the induction time was constant, coincides with the homogenous zone identified earlier.

Table 10: Relation between the initial pH and SI for $CaCO_3$ phases at 50% recovery in the presence of Mg^{2+} only or Mg^{2+} and SO_4^{2-}

	Initial pH	t_{nd} min	Log t_{nd}	SI					
				Calcite	Aragonite	Vaterite	Monoh-ydrated	Hexah-ydrated	Amorph-ous
Mg+SO4_50% R	8.18	1296	3.1	1.33	1.19	0.75	-0.07	-0.42	-0.77
	8.25	927	3.0	1.37	1.23	0.79	-0.03	-0.38	-0.73
	8.49	602	2.8	1.52	1.38	0.94	0.12	-0.23	-0.58
	8.53	692	2.8	1.54	1.40	0.96	0.14	-0.21	-0.56
	8.75	412	2.6	1.65	1.51	1.07	0.25	-0.10	-0.45
	9.03	202	2.3	1.74	1.60	1.16	0.34	-0.01	-0.36
	9.11	220	2.3	1.76	1.62	1.18	0.36	0.01	-0.34
	9.15	267	2.4	1.77	1.63	1.19	0.37	0.02	-0.33
	9.34	262	2.4	1.80	1.66	1.22	0.40	0.05	-0.30
	9.59	272	2.4	1.82	1.68	1.24	0.42	0.07	-0.28
	9.67	321	2.5	1.82	1.68	1.24	0.42	0.07	-0.28
	9.78	210	2.3	1.82	1.68	1.24	0.42	0.07	-0.28
	9.8	376	2.6	1.82	1.68	1.24	0.42	0.07	-0.28
	9.9	299	2.5	1.82	1.68	1.24	0.42	0.07	-0.28
Mg_50% R	8.19	870	2.9	1.38	1.24	0.80	-0.02	-0.37	-0.72
	8.51	460	2.7	1.57	1.43	0.99	0.17	-0.18	-0.53
	8.53	450	2.7	1.58	1.44	1.00	0.18	-0.17	-0.52
	8.60	302	2.5	1.62	1.48	1.04	0.22	-0.13	-0.48
	8.80	268	2.4	1.69	1.55	1.11	0.29	-0.06	-0.41
	8.86	182	2.3	1.71	1.57	1.13	0.31	-0.04	-0.39
	9.09	280	2.4	1.77	1.63	1.19	0.37	0.02	-0.33
	9.37	252	2.4	1.81	1.67	1.23	0.41	0.06	-0.29
	9.77	272	2.4	1.83	1.69	1.25	0.43	0.08	-0.27
	9.81	260	2.4	1.83	1.69	1.25	0.43	0.08	-0.27

In this research the values obtained in the identified homogenous range in the presence of Mg^{2+} or Mg^{2+} and SO_4^{2-} are 250-300 min. In same identified homogenous range the induction time for the control solution was less that 1 min (lower than identified pH measurement interval). This may suggest that the presence of Mg^{2+} or Mg^{2+} and SO_4^{2-} increases the induction time by at least 250 times compared to the control solution. These results are comparable to that obtained by Sohnel and Mullin, 1982 [37], where they also claimed that the presence of magnesium increases the induction time by about 100 times compared to control solutions.

Stable CaCO₃ phase

Figure 17: ESEM pictures for the final crystal phases in the presence of only SO_4^{2-} (top left) and only Mg^{2+} (top right and middle right) and in the presence of Mg^{2+} and SO_4^{2-} (middle left and lower left and right)

Eight wet samples were taken from the experimental reactors at the end of representative experiments (2 days) for recovery of 30% and 50%. The samples were rinsed with ultra-pure water before being stored for 5 months for aging and ripening before analysis ESEM and XRD. The results from ESEM showed mushroom and star shape crystals of aragonite in experiments containing Mg^{2+} or Mg^{2+} and SO_4^{2-}. On the contrary distinct cubic crystal calcite shapes were found in the absence of Mg^{2+} (Figure 17).

The results from the XRD analysis confirmed the results which were obtained using the ESEM. The results shown in Figures 18-20 illustrate that aragonite is the only present $CaCO_3$ species in the presence of Mg^{2+}. On the other hand, calcite was the dominant species in cases of experiments with only SO_4^{2-}. These results suggest that the Mg^{2+} not only delays or inhibits the nucleation and the subsequent growth of $CaCO_3$ but it also prevents the formation of calcite even after 5 months of ripening and therefore acts as a poison agent for the calcite formation. These results are consistent with literature results obtained for $CaCO_3$ crystals in the presence of Mg^{2+} [37].

Although these results do not represent the first formed $CaCO_3$ carbonate phase, it is an indication that using the solubility of calcite in seawater medium containing Mg^{2+} may be misleading.

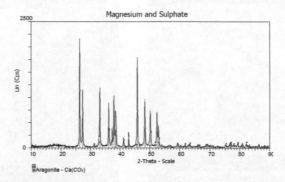

Figure 18: XRD Analysis for the crystal final phases in the presence of Mg^{2+} and SO_4^{2-}

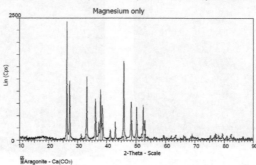

Figure 19: XRD Analysis for the crystal final phases in the presence of only Mg^{2+}

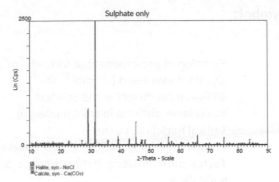

Figure 20: XRD Analysis for the crystal final phases in the presence of only SO_4^{2-}

6.6   Conclusions

- The induction time of supersaturated synthetic solutions of calcium carbonate simulating SWRO concentrates for all the experimental sets of 30% and 50% recovery decreased by increasing the initial experimental pH and the corresponding SI.

- When adding SO_4^{2-}, Mg^{2+} or Mg^{2+} and SO_4^{2-} to synthetic SWRO concentrate, the induction time increased by 3.3, 4.5 and 6.6 times, respectively, for the recovery of 30%, compared to that obtained In the absence of Mg^{2+} and SO_4^{2-} at the starting pH of 8.3. Similar results were obtained for 50% recovery as the induction time increased by 11, 28 and 39 times at a starting pH of 8.3 in the presence of SO_4^{2-}, Mg^{2+} or Mg^{2+} and SO_4^{2-}, respectively. This increase of the induction time in that order is due to the presence of Mg^{2+} and SO_4^{2-} which actually retard the crystal growth of $CaCO_3$.

- Calculating the activity coefficients for calcium and carbonate showed that the increase in induction time due to the presence of magnesium and sulphate is due to growth

- Crystal sample analysis using ESEM and XRD showed aragonite crystal as the only calcium carbonate phase formed after 5 months of ripening and aging in the case of 'Mg^{2+} and SO_4^{2-}' and 'only Mg^{2+}' experimental sets, while in the case of 'only SO_4^{2-}' experimental sets, it was found that the calcite crystal phase was the most stable final phase of the experiment. This indicates that magnesium may play an important role in the crystallization mechanism of calcium carbonate through the prevention of calcite formation.

6.7   List of symbols

A	Function of pre-exponential factor ($s^{-1}m^{-3}$)
B	Constant expressed ($L^{3/2}mol^{-1/2}nm^{-1}$)
D	Diffusion coefficient in the solution
d	Interplaner distance in solid phase (m)
$r*$	Critical nuclei radius (m)
g	Formal order of growth
$f(\theta)$	Factor differentiating heterogeneous and homogenous nucleation
I	Ionic strength (mol/L)
IAP	Ionic activity product (mol^2/L^2)
J_s	Steady state nucleation rate (nuclei/min/cm^3)
k	constant in the induction time equation
k_b	Boltzmann constant (J/K)
k_g	Kinetic constant of crystal growth
K_{sp}	Solubility product ($mole^2/L^2$)
K_{so}	Solubility at standard conditions ($mole^2/L^2$)
N_A	Avogadros number ($mole^{-1}$)
P	Pressure (psi)
S_a	Supersaturation ratio
SI	Supersaturation Index
T	Absolute temperature in Kelvin
t_{ind}	Induction time in minutes unless mentioned otherwise (min)
t_n	Nucleation time (min)
t_g	Growth time (min)
v	Number of ions into which a molecule dissociates
ϑ	Molecular volume (cm^3/mole)
x	Number of building units arriving on a unit surface of nucleus in unit time ($m^{-1}.sec^{-1}$)

Greek letters

β	Geometric factor for nucleation
β'	Geometric factor for growth
γ_+	Cation activity coefficient
γ_-	Anion activity coefficient
γ_s	Surface energy (J/m^2)
Ω	Pre-exponential factor in the nucleation rate equation ($s^{-1}m^{-3}$)
$\varnothing$	Precipitation driving force (J)

6.8   References

1.   Wilf, M., The guidebook to membrane desalination technology. 2007, L'Aquila: Desalination publications.

2.   Sohnel, O. and J. Garside, Precipitation basis principals and industrial applications. 1992, Oxford: Butterworth-Heinemann.

3.   Dove, P.M., J.J. De Yoreo, K.J. Davis, Inhibition of $CaCO_3$ crystallization by small molecules: the magnesium example, in In From Fluid-Solid Interfaces to Nanostructural Engineering Vol II: Assembly in Hybrid and Biological Systems, J.J.D.Y.a.X.Y. Liu, Editor. 2004, Plenum/Kluwer Academic Press.: New York. p. 55-82.

4.   Sheikholeslami, R., Assessment of the scaling potential for sparingly soluble salts in RO and NF units. Desalination, 2004. 167: p. 247-256.

5.   Treccani, L., Protein Mineral Interaction of Purified Nacre Proteins with Carbonate Crystals. 2006, University of Bremen: Bremen.

6.   Chong, T.H. and R. Sheikholeslami, Thermodynamics and kinetics for mixed calcium carbonate and calcium sulfate precipitation. Chemical Engineering Science, 2001. 56(18): p. 5391-5400.

7.   Loste, E., et al., The role of magnesium in stabilizing amorphous calcium carbonate and controlling calcite morphologies. Journal of Crystal Growth 2003. 254: p. 206-218.

8.   Waly, T., et al., Reducing the calcite scaling risk in SWRO: role of Mg^{2+} & SO_4^{2-}, in IDA world congress: Desalination for a better world. 2009: Dubai.

9.   Koutsoukos, P. and C. Kontoyannis, Precipitation of calcium carbonate in aqueous solutions. J. Chem. Soc., Faraday Trans.I, 1984. 80: p. 1181-1192.

10.   Koutsoukos, P.G. and C.G. Kontoyannis, Prevention and inhibition of calcium carbonate scale. Journal of Crystal Growth, 1984. 69: p. 367-376.

11.   Liu, S.-T. and G.H. Nancollas, A kinetic and morphological study of the seeded growth of calcium sulfate dihydrate in the presence of additives. Journal of Colloid and Interface Science, 1975. 52(3): p. 593-601.

12.   Liu, S.-T. and G.H. Nancollas, The kinetics of crystallization and dissolution of barium oxalate dihydrate and cadmium oxalate trihydrate. Journal of Inorganic and Nuclear Chemistry, 1976. 38(3): p. 515-522.

13.   Liu, S.-T. and G.H. Nancollas, The crystal growth and dissolution of barium sulfate in the presence of additives. Journal of Colloid and Interface Science, 1975. 52(3): p. 582-592.

14.   Roelands, C.P.M., Polymorphism in Precipitation Processes. 2005, TU Delft: Delft.

15.   Sabbides, T.G. and P.G. Koutsoukos, The crystallization of calcium carbonate in artificial seawater; role of the substrate. Journal of Crystal Growth 1993. 133: p. 13-22.

16.   Snoeyink, V. and D. Jenkins, Water Chemistry. 1980, New York: John Wiley & Sons, Inc.

17.   Elfil, H. and H. Roques, Role of hydrate phases of calcium carbonate on the scaling phenomenon. Desalination, 2001. 137: p. 177-186.

18.   Königsberger, E., J. Bugajski, and H. Gamsjäger, Solid-solute phase equilibria in aqueous solution: II. A potentiometric study of the aragonite-calcite transition. Geochimica et Cosmochimica Acta, 1989. 53(11).

19.   Lippmann, F., Sedimentary carbonate minerals. 1973, Berlin: Springer-Verlang.

20.   Voinescu, A.E., Biomimetic Formation of $CaCO_3$ Particles Showing Single and Hierarchical Structures, in Chemical and Pharmaceutical. 2008, University of Regensburg Regensburg

21.   Ohtaki, H., Crystallization processes. 1998, New York: John Wiley & Sons

22.   Elfil, H. and H. Roques, Kinetics of the precipitation of calcium sulfate dihydrate in a desalination unit. Desalination 2003. 157: p. 9-16.

23.   Mindat.org. Calcium Carbonate polymorphs. http://www.mindat.org 2010.

24.   Bischoff, J.L. and K.S. Pitzer, Phase relations and adiabats in boiling seafloor geothermal systems. Earth and Planetary Science Letters, 1985. 75(4): p. 327-338.

25.   Abdel-Aal, E., M. Rashad, and H. El-Shall, Crystallization of calcium sulfate dihydrate at different supersaturation ratios and different free sulfate concentrations. Cryst. Res. Technol., 2004. 39: p. 313–321.

26.   Chen, T., A. Neville, and M. Yuan, Assessing the effect of Mg^{2+} on $CaCO_3$ scale formation–bulk precipitation and surface deposition. Journal of Crystal Growth, 2005. 275: p. 1341-1347.

27.   Gledhill, D.K. and J.W. Morse, Calcite solubility in Na-Ca-Mg-Cl brines. Chemical Geology, 2006. 233: p. 249-256.

28.   Gledhill, D.K. and J.W. Morse, Calcite dissolution kinetics in Na-Ca-Mg-Cl brines. Geochimica et Cosmochimica Acta, 2006. 70: p. 5802-5813.

29.   Ingle, S.E., Solubility of calcite in the ocean. Marine Chemistry, 1975. 3: p. 301-319.

30.   Mustafa, G.M., The study of pretreatment options for composite fouling of reverse osmosis membranes used in water treatment and production, in Chemical science and Engineering. 2007, University of New South Wales.

31.   Slack, J.G., Calcium carbonate hexahydrate: its properties and formation in lime-soda softening. Water Research, 1980. 14: p. 799-804.

32.   Zuddas, P. and A. Mucci, Kinetics of calcite precipitation from seawater: I. A classical chemical kinetics description for strong electrolyte solutions. Geochimica et Cosmochimica Acta, 1994. 58(20): p. 4353-4362.

33.   Zuddas, P. and A. Mucci, Kinetics of calcite precipitation from seawater: II the influence of the ionic strength. Geochimica et Cosmochimica Acta, 1998. 62: p. 757-766.

34.   Zuddas, P., K. Pachana, and D. Faivre, The influence of dissolved humic acids on the kinetics of calcite precipitation from seawater solutions. Chemical Geology, 2003. 201: p. 91– 101.

35.   Van Der Weijden, R., Interactions between Cadmium and Calcite. 1995, Utrecht University: Utrecht.

36.   Curti, E., Coprecipitation of radionuclides: basic concepts, literature and first applications. 1997, Paul Scherrer Institute: Villigen, Switzerland.

37.   Sohnel, O. and J.W. Mullin, Precipitation of calcium carbonate. Journal of Crystal Growth, 1982. 60: p. 239-250.

38.   Gower, L.B. and D.J. Odom, Deposition of calcium carbonate films by a polymer-induced liquid-precursor (PILP) process. Journal of Crystal Growth, 2000. 210: p. 719-734.

39.   Abbona, F., H.E.L. Madsen, and R. Boistelle, The initial phases of calcium and magnesium phosphates precipitated from solutions of high to medium concentrations. Journal of Crystal Growth, 1986. 74: p. 581-590.

40.   Sohnel, O., Electrolyte crystal-aqueous solution interfacial tensions from crystallization data. Journal of Crystal Growth, 1982. 57: p. 101-108.

41.   Morse, J.W., Q. Wang, and M.Y. Tsio, Influences of temperature and Mg:Ca ratio on $CaCO_3$ precipitates from seawater. Geology, 1997. 25: p. 85-96.

42.   Burton, E.A. and L.M. Walter, The effects of PCO_2 and temperature on magnesium incorporation in calcite in seawater and $MgCl_2$-$CaCl_2$ solutions. Geochim. Cosmochim. Acta, 1985. 49: p. 713-722.

43.   Mucci, A., Influence of temperature on the composition of magnesian calcite overgrowths precipitated from seawater. Geochim. Cosmochim. Acta, 1987. 51: p. 1977-1991.

44.   Pokrovsky, O.S., Precipitation of calcium and magnesium carbonates from homogeneous supersaturated solutions. Journal of Crystal Growth, 1998. 186: p. 233-244.

45.   Mucci, A., R. Canuel, and S. Zhong, The solubility of calcite and aragonite in sulfate free seawater and the seeded growth kinetics and composition of the precipitates at 25 °C. Chemical Geology, 1989. 74: p. 309-320.

46.   Drioli, E., et al., Integrated system for recovery of CaCO3, NaCl and $MgSO_4 \cdot 7H_2O$ from nanofiltration retentate. Journal of Membrane Science, 2004. 239: p. 27-38.

47.   Appelo, C. and D. Postma, Geochemistry, groundwater and pollution. 2nd ed. 2005, Leiden: A.A.Balkema publishers.

48.   Hardie, L.A., Dolomitization; a critical view of some current views. Journal of sedimentary research, 1987. 57: p. 166-183.

49.   Mullin, J.W., Crystalization. 4th. ed. 2001, Oxford: Butterworth Heinemann.

50.   Golubev, S.V., O.S. Pokrovsky, and V.S. Savenko, Unseeded precipitation of calcium and magnesium phosphates from modified seawater solutions. Journal of Crystal Growth, 1999. 205: p. 354-360.

51.   Madsen, H.E.L., Influence of foreign metal ions on crystal growth and morphology of brushite ($CaHPO_4$, $2H_2O$) and its transformation to octacalcium phosphate and apatite. Journal of Crystal Growth, 2008. 310: p. 2602-2612.

52.   Le Corre, K.S., et al., Impact of calcium on struvite crystal size, shape and purity. Journal of Crystal Growth, 2005. 283: p. 514-522.

53.   Yu, H., The mechanism of composite fouling in Australian sugar mill evaporators by calcium oxalate and amorphous silica, in Chemical Engineering and Industrial chemistry. 2003, University of New South Wales: Sydney.

54.   Bansal, B., H. Muller-Steinhagen, and X.D. Chen, Effect of suspended particles on crystallization fouling in plate heat exchangers. ASME, 1997. 119: p. 568-574.

55.   Sheikholeslami, R., Mixed salts--scaling limits and propensity. Desalination, 2003. 154(2): p. 117-127.

56.   Turek, M. and P. Dydo, Electrodialysis reversal of calcium sulphate and calcium carbonate supersaturated solution. Desalination, 2003. 158: p. 91-94.

57.   Schippers, J., Desalination methods. 2003, Delft: UNESCO-IHE.

58.   Boerlage, S., Scaling and particulate fouling in membrane filtration system, in Sanitary Engineering. 2002, IHE: Delft.

59.   Jiang, S., Crystallization Kinetics in Polymorphic Organic Compounds. 2009, TUDelft: Delft.

60.   Van Der Leeden, M., The Role of Polyelectrolytes in Barium Sulphate Precipitation. 1991, TUDelft: Delft.

61.   Patel, S. and M. Finan, New antifoulants for deposit control in MSF and MED plants. Desalination, 1999. 124: p. 63-74.

62.   Lee, S., J.S. Choi, and Z.H. Lee, Behaviors of dissolved organic matter in membrane desalination. Desalination, 2009. 238: p. 109-116.

63.   Butler, J., Carbon Dioxide equilibria and their applications. second ed. 1982, California: Addison-Wesley.

64.   Sohnel, O. and J. Garside, On Supersaturation Evaluation For Solution Growth. Journal of Crystal Growth, 1981. 54: p. 358-360.

65.   Bromley, L.A., Thermodynamic properties of strong electrolytes. AIChE Journal, 1973. 19(2): p. 313-320.

66.   Bromley, L.A., Approximate individual ion values of β (or B) in extended Debye-Hückel theory for uni-univalent aqueous solutions at 298.15 °K. J. Chem. Thermodynamics, 1972. 4: p. 669-673.

67.   Pitzer, K.S. and G. Mayorga, Thermodynamics of electrolytes (III) Activity and osmotic coefficients for 2-2 electrolytes. J. of Soln. Chem., 1973. 3: p. 539-546.

68.   Pitzer, K.S., Thermodynamics of electrolytes. I. Theoretical basis and general equations. J. Phys. Chem., 1973. 77: p. 268-277.

69.   ASTM, Calculation and Adjustment of the Stiff and Davis Stability Index for Reverse Osmosis. 2001, ASTM International: West Conshohocken, PA, United States.

70.   Wang, Y., Composite fouling of calcium sulphate and calcium carbonate in a dynamic seawater reverse osmosis unit. 2005, University of New South Wales: Sydney.

71.   Darton, E., Membrane chemical research: centuries apart. Desalination, 2000. 132: p. 121-131.

72.   Sorber, A. and R. Valenzuela, Evaluation of an electrolytic water conditioning device for the elimination of water-formed scale deposits in domestic water systems, in center for research in water resources. 1982, Texas university: Austin.

73.   Sohnel, O. and J.W. Mullin, Interpretation of Crystallization Induction Periods. Journal of Colloid and Interface Science, 1988. 123(1): p. 43-50.

74.   Sohnel, O. and J.W. Mullin, The role of time in metastable zone width determination. Chemical Eng. Res. Des., 1988. 66: p. 537-540.

75.   Sohnel, O. and J.W. Mullin, Influence of mixing on batch precipitation Crystal Research and Technology 1987. 22(10): p. 1235 - 1240.

76.   Avrami, M., Kinetics of phase change: I general theory. Journal of chemical physics, 1939. 7(12): p. 1103-1112.

77.   Avrami, M., Kinetics of phase change: II transformation time relations for random distribution of nuclei. Journal of chemical physics, 1940. 8(2): p. 212-224.

78.   Avrami, M., Kinetics of phase change: III granulation, phase change, and microstructure. Journal of chemical physics, 1941. 9(2): p. 177-184.

79.   Sergei, V., O.S. Pokrovsky, and V.S. Savenko, Unseeded precipitation of calcium and magnesium phosphates from modified seawater solutions. Journal of Crystal Growth, 1999. 205: p. 354-360.

80.   Golubev, S.V., O.S. Pokrovsky, and V.S. Savenko, Homogeneous precipitation of magnesium phosphates from seawater solutions. Journal of Crystal Growth, 2001. 223: p. 550-556.

Chapter 7

Conclusions and recommendations

7.1   Conclusions

This research tried to illustrate that the seawater environment is unique in its characteristics and composition. Therefore the description of seawater scaling behaviour as a function of only Ca^{2+} and HCO_3^- is most likely not enough to describe the behaviour of calcium carbonate in SWRO systems. The magnesium and sulphate in seawater, in addition to the complexity of the mechanism of nucleation and growth, encourage the need for further research to determine the real saturation of $CaCO_3$ in SWRO concentrates and to understand the kinetics of its precipitation.

The first step of this research was to develop a method capable of measuring the induction time in a reproducible yet accurate way. Three methods have been evaluated namely: measurement of calcium with ICP, conductivity and pH.

Two different methods, making use of ICP were investigated. In the first method the samples were filtered through a filter with pores of 0.2 um and calcium was measured in the filtrate with ICP. The second was to measure the crystals retained after filtering of the sample, after which, the filters were acidified and the calcium concentration measured with ICP. Both methods turned out to be rather inaccurate.
Conductivity measurements showed to be inaccurate as well, since the change in conductivity due to precipitation of calcium carbonate was too low to be detected accurately.
Measurements with a new (on the market) and highly sensitive and stable pH meter, were very accurate.
No further effort was spent on optimizing the methods to measure Ca^{2+} via ICP and pH measurements were employed in all induction time experiments.
In addition the effect carbon dioxide exchange with the atmosphere has been investigated, since literature does not give a clear answer on the question; whether a reactor should be open or closed. The outcome of the investigations is that open reactors showed obviously a significant exchange of carbon dioxide with the atmosphere resulting in large deviations in observed induction times in closed reactors. Consequently closed reactors were applied in this study.

In the second part of this study, the effect foreign of particles on the crystallization process of $CaCO_3$ in seawater was simulated and the effect of stirring speed in the reactors determined. Synthetic supersaturated SWRO concentrate (recovery of 50%) was produced by dissolving pure salts (99.9%) (sodium, chloride and hydrogen carbonate only) in ultra-pure water to avoid the presence of foreign particles in the supersaturated solution. In addition the supersaturated solutions were pre-filtered with either 0.2 μm; 0.1 μm and 100 kDa (0.03 μm) filters to remove any particles present in the solution.

No pronounced effect of pre-filtration with 0.2μm; 0.1μm and 100 kDa (0.03 μm) was observed on the induction time for the synthetic SWRO concentrate with initial S&DSI of 0.12 to 1.28. A possible explanation for this result is that there are no particles in the synthetic solutions to affect the induction time since ultra-pure water was used in its preparation or induction time is only affected by particles smaller than 100kDa. In this study, different glass nano-particles concentrations were dosed to determine the effect of particles numbers on the induction time. Three doses were used $6.4*10^{10}$ particle/L, $8*10^{12}$ particle/L and $6.4*10^{13}$ particle/L where the first represents the amount found in real seawater (10^7-10^{10} particle/L). These glass particles added 123%, 16% and 1% to the surface area of the glass reactor wall.

The results showed immediate precipitation when the solution was subjected to a high suspension glass particles ($6.4*10^{13}$ particle/L). In such case, induction time could not be measured. In the case of a particle concentration of $8*10^{12}$ /L, the induction time was 59 minutes and 78 minutes for an initial pH of 8 and 7.93, respectively. When compared to induction times measured for (filtered) solutions, these results are 30% lower than the values of 90 and 120 minutes. Finally the induction time showed slight/no decrease for the lower particle dose ($6.4*10^{10}$ particle/L) which is the range of particle numbers present in seawater.

Different stirring speeds had no effect on the determined induction times.

The third studied aspect was to determine deficiencies in saturation indices used for $CaCO_3$. The S&DSI, which is based on experimental synthetic solutions containing sodium, chloride, hydrogen carbonate and calcium, is widely used for the saturation calculation of $CaCO_3$. Unlike other indices, e.g., SI and S_a, the solubility and dissociation constants used in the S&DSI are not known. In order to elucidate the crystallizing phase incorporated in the S&DSI, The SI and the S&DSI were compared at the same pH levels. The SI calculations were performed for different 6 phases of $CaCO_3$ for low and high ionic strength water (I = 0.054, 1.12, 1.34, 1.61 mole/L). Activity coefficients, to take into account the effect of salinity on the SI were calculated using Phreeqc. Results showed that for low ionic strength solution (I=0.054 mole/L), the S&DSI matched the SI when the solubility of calcite was incorporated into the SI calculations. On the contrary, for high ionic strength water (I = 1.12, 1.34, 1.61 mole/L) the S&DSI matched the SI using the solubility of vaterite in its calculation. The results suggested that vaterite and not calcite maybe the precipitating phase in seawater. Incorporating the solubility of vaterite and not calcite in the SI for seawater calculation will result in a decrease in SI by 0.5 units compared to that using the solubility of calcite. These results were confirmed with microscopic analysis of the formed crystals at the end of the induction time experiments (24hrs) as vaterite was found in the solution.

The mechanism of nucleation is known to be related to the degree of supersaturation. In seawater reverse osmosis systems, the membrane may be subjected to different levels of supersaturation and therefore the mechanism of nucleation may change. For the purpose of determining the mechanism of nucleation expected to be involved for $CaCO_3$ in SWRO system, the relation between $Log\ t_{ind}$ and $T^{-3}Log^{-2}(S_a)$ was implemented. Three different slopes were found where the highest slope represents the homogenous nucleation and the lowest line represents the heterogeneous nucleation zone. An intermediate zone with an intermediate line slope, and consequently a different surface energy was recognized. The nucleation mechanism involved in the induction time experiment was closely related to the initial synthetic solution saturation. Homogenous nucleation predominates when the initial solution saturation exceeds that of hexahydrated $CaCO_3$ (Ikaite) and heterogeneous nucleation predominates when the initial saturation is less than the monohydrated, but higher than the vaterite. Finally, the area in between with the intermediate slope was created when the initial saturation is below the Ikaite but higher than the hexahydrated. Results suggested that heterogeneous nucleation is the mechanism expected to be encountered in SWRO systems.

The fourth part of this research was dedicated to the study of the concentrate pH in SWRO plants as the real saturation and consequently, the acid and/or antiscalants dosing for preventing $CaCO_3$ precipitation cannot be determined without accurate prediction of the concentrate pH. Meanwhile, accurate determination of the concentrate pH needs to take the acidity constants for seawater and the rejection values of HCO_3^- and CO_3^{2-} into consideration. Results suggest that pH calculations using the CO_2-HCO_3^--CO_3^{2-} system equilibrium equations, simulations using two commercial software programs from membrane suppliers and also the evaporation software package Phreeqc, was not able to simulate the real concentrate pH measured in a SWRO plant using the North seawater. Field measurements, in a pilot plant indicate that the pH of the concentrate tends to be lower than that of the feed if the feed pH is higher than 7.09 and tends to be lower if the feed pH is lower than 7.09. Furthermore, Results suggest that at pH lower than 7.0, the equilibrium between hydrogen carbonate/carbon dioxide dominates the pH in the concentrate while at pH levels higher than 8.0, the equilibrium hydrogen carbonate/ carbonate dominates.

On the other hand, results suggest that using the dissociation constants corrected for synthetic NaCl solutions resulted in pH values up to 0.05- 0.2 pH units compared to dissociation constants for real seawater compositions. The later, can predict the concentrate pH closer to the real value compared to the equations, is close to the real concentrate pH compared to that using dissociation constants corrected for synthetic NaCl solutions.
Using different rejection values for the HCO_3^- and CO_3^{2-} of 90 and 99% (respectively) only affected the predicted pH by 0.02 pH units when compared to the assumption of 100% rejection. These results show that the pH prediction of SWRO concentrate is more sensitive to the effect of ionic strength on the activities of the ions involved than to different rejection ratios.

Seawater consists of various organic and inorganic ions which may affect the saturation and the crystallization of $CaCO_3$. In the last part of this research, the presence of inorganic ions namely, Mg^{2+} and SO_4^{2-} in seawater and their effect on the crystallization mechanism of $CaCO_3$ was studied. The induction time of supersaturated synthetic solutions of calcium carbonate simulating SWRO concentrates was increased by 40-200 times in the presence of SO_4^{2-}, Mg^{2+} or Mg^{2+} and SO_4^{2-} compared to that in the absence of Mg^{2+} and SO_4^{2-} at the same initial pH. This suggests that studying $CaCO_3$ scaling in SWRO plants without taking into account the role of Mg^{2+} and SO_4^{2-} ions may result in inaccurate scaling prediction, based on induction time measurements and hence an overestimation the probability of scaling and consequently, the acid and/or antiscalant doses.

ESEM and XRD showed aragonite crystal as the only calcium carbonate phase formed after 5 months of aging in the case of 'Mg^{2+} and SO_4^{2-}'and 'only Mg^{2+}' containing experiments solutions, while in the case of 'only SO_4^{2-}' containing solutions, it was found that calcite was the stable phase after 5 months of the experiment. This indicates that magnesium may play an important role in the crystallization mechanism of calcium carbonate through the prevention of calcite formation.

Finally, the outcome of this study indicates that the pH of concentrates of seawater reverse osmosis plants are lower than commonly expected. This effect is attributed to the effect of ionic strength on the activities of the ions involved and the mechanisms governing the pH in the feed – concentrate. As a consequence, the degree of supersaturation is lower as well.

If vaterite is governing the solubility of calcium carbonate – as indicated in induction time measurements in artificial seawater solutions, containing magnesium and sulphate as well - the saturation of seawater concentrate is expected to be nearly 0.7 SI units lower.

This opens, together with the rather long induction time measured in artificial seawater with magnesium and sulphate the opportunity to reduce or even stop dosing acid/antiscalant used to prevent $CaCO_3$ scaling in SWRO plants. To prove this hypothesis, a SWRO pilot plant in the Netherlands has operated on open seawater at 40% recovery for more than 6 months without any acid/antiscalant and showing no scaling.

7.2   Recommendations for future work

Although, this research was able to answer several questions raised. Future research could be a fruitful subject for one or more PhD studies in this area if some or all of the following recommendations are considered.

The induction time measurements proved themselves to be valuable in describing the crystallization of $CaCO_3$ SWRO concentrates in the presence of different factors, namely salinity, and the presence of organic and inorganic ions in the water. Maximizing the benefit of this tool requires a link between field and laboratory induction time measurements, in order to determine the safe induction time before which scaling will probably not occur in the SWRO plants. This could be established by installing a "in-line scaling monitor". This monitor measures the development of the normalized flux in the last element in the pressure vessel.

The standardization of the induction time measurement procedure and apparatus are key factors in having reliable and consistent measurements. This standardization should take into account, e.g., the reactor surface area, volume and material type.

Although this research looked widely in the literature to find methods used in induction time measurements, there is a need to investigate other methods, e.g., laser turbidity measurements or online photometric measurements. The faster the detection of the first crystals, the more reliable the induction time measurements are and the more it represents nucleation rather than growth.

In this research induction time experiments we usually conducted at $20^{\circ}C$. It might be useful to do these experiments at a broad range of temperatures, e.g., $0^{\circ}C$ to $35^{\circ}C$

Natural organic matter was reported to affect the mechanism of $CaCO_3$ nucleation. However, it is not yet well known to what extend organic matter is influencing the induction time and which part of NOM affects the formation of $CaCO_3$. Future investigation could take into consideration the simulation of different NOM fractions, e.g, humic substances, to investigate which is more influential on the $CaCO_3$ nucleation and crystal growth.

Further pilot testing on RO plant concentrate is recommended to be able to verify the results obtained in this research and for further fine tuning of earlier conclusions.

PUBLICATIONS

1. Waly, T.; Saleh, S.; Kennedy, M.D.; Witkamp, G.J.; Amy, G. and Schippers, J.C. Will calcium carbonate really scale in seawater reverse osmosis? In Proceeding of the EDS (Ed.) EuroMed Conference, 2008. Dead Sea, Jordan.

2. Waly, T.; Saleh, S.; Kennedy, M.D.; Witkamp, G.J.; Amy, G. and Schippers, J.C. Will calcium carbonate really scale in seawater reverse osmosis? Desalination and Water Treatment, 2009. 5 p252-256

3. Waly, T.; Munoz, R.; Kennedy, M.D.; Witkamp, G.J.; Amy, G. and Schippers, J.C. Role of particles on Calcium carbonate scaling of SWRO systems. In proceeding of the EDS (Ed.) Desalination for the Environment Conference. 2009. Baden-Baden, Germany

4. Waly, T.; Saleh, S.; Kennedy, M.D.; Witkamp, G.J.; Amy, G. and Schippers, J.C. Reducing the calcite scaling risk in SWRO: role of Mg^{2+} & SO_4^{2-}. In proceeding of the IDA World Congress. 2009. Dubai, UAE

5. Waly, T.; Munoz, R.; Kennedy, M.D.; Witkamp, G.J.; Amy, G. and Schippers, J.C. Role of particles on calcium carbonate scaling of SWRO systems. Desalination and Water Treatment, 2010. 18 p103-111.

6. Munoz, R., Schippers, J.C.; Amy, G. Kennedy, M.D.; Waly, T. and Witkamp, G.J Impact of the ionic strength on the estimation of the pH of the concentrate in Sea water reverse osmosis (SWRO). In Proceeding of the 2nd International Congress on Water Management in the Mining Industry. 2010. Santiago, Chile

7. Waly, T.; Kennedy, M.D.; Witkamp, G.J.; Amy, G. and Schippers, J.C. Predicting $CaCO_3$ scaling: Towards a correct pH calculation in SWRO concentrates. In Proceedings of the MDIW membranes in drinking and industrial water treatment world congress. 2010. Trondheim, Norway

8. Waly, T.; Kennedy, M.D.; Witkamp, G.J.; Amy, G. and Schippers, J.C. Predicting $CaCO_3$ scaling: Towards a correct pH calculation in SWRO concentrates. Desalination, submitted

9. Waly, T.; Kennedy, M.D.; Witkamp, G.J.; Amy, G. and Schippers, J. On the induction time of $CaCO_3$ in high ionic strength synthetic seawater. Desalination and Water Treatment, accepted

10. Waly, T.; Kennedy, M.D.; Witkamp, G.J.; Amy, G. and Schippers, J.C. The role of inorganic ions in the calcium carbonate scaling of seawater reverse osmosis systems. Desalination, submitted

CURRICULUM VITAE

Of Tarek Waly, born in Cairo, Egypt on 10$^{th.}$ of May 1974

Education

❑ Ph.D. (Urban water & Infrastructure) TU Delft University and UNESCO-IHE Delft, The Netherlands (2011)

❑ M. Sc. (Sanitary Engineering), UNESCO-IHE Delft, The Netherlands (2004)

❑ B.Sc. (Civil Engineering) Ain Shams University, Egypt (1997).

Key qualifications

MAIN DISCIPLINE   Water & Waste Water Treatment

SPECIALIZATION

❑ Desalination systems Design, Modelling and Optimization
❑ Domestic and Industrial water, wastewater treatment systems - Design, Modelling, Diagnosis and Control
❑ Water management and water cycle
❑ Research and Development
❑ Consultancy

Main projects

❑ Optimization of Seawater Reverse Osmosis desalination plant system using North sea water as raw water source and located in the provinces of Zeeland, The Netherlands. The aim of this project to run a SWRO plant using the North sea water as the feed water without acid or antiscalants addition.

❑ Take part in the design and optimization of 3 brackish water reverse osmosis units for the gas and petroleum companies of Agiba and Khalda in the west desert, Egypt. The three plants produced water for drinking and industrial use from high salinity brackish water with elevated temperature.

❑ Take part in the conceptional design and tender documents preparation of

3 brackish water reverse osmosis units for residential used in the El Arish, Egypt. The three plants produced water for drinking and industrial use with high brackish water with elevated temperature as feed water.

❑ Conduct a feasibility study and plant design of 20,000 m³/d SWRO plant for the cultural village project in Doha, Qatar for commercial and residential use including cooling water. The feasibility incorporated the use of solar energy for power generation in the desalination plant.

❑ Conduct the conceptional design and tender documents preparation for 8 water treatment plants using water with high levels of iron & Manganese in Elwadi Elgadid governorate, Egypt.

❑ Conduct the conceptional design for SWRO using red sea water as the feed water Shalateen, Egypt.

❑ Take part in the conceptional design and pricing of the rehabilitation and upgrading of 6 SWRO plants on the red sea cost, Egypt.

❑ Take part in the conceptional design and pricing of the rehabilitation and upgrading of 2 waste water treatment plants using flat sheet MBR on Garbeya, Egypt.

❑ Conduct conceptional design for 20 small and medium water and waste water plants for domestic use in Egypt.

❑ Take part in the conceptional design and pricing of 6 waste water treatment plants using RBC technique, Alexandria, Egypt.

❑ Take part in the conceptional design and pricing of 3 mega scale water treatment plants using river Nile water as the source water in Maadi, Roud el Farag and Imbaba , Egypt

❑ Take part in the conceptional design and pricing of industrial waste and water treatment plants for industrial applications of power generation, copper wiring, cosmetics, Dairy and petrochemical industries located on various industrial parks, Egypt

❑ Design, optimize and model the use of ion exchange for effective low molecular weight NOM particles and its effect on the energy requirements for UV disinfection and UF fouling potential. The work was done on a pilot scale with Isselmeer (Lake) water as the source water. The work result was the base for constructing the world largest UV disinfection plant in Andijke, the Netherlands.

❑ Develop a monitoring protocol for the prediction of scaling in SWRO systems as a part of a European Union project (EU-Medina) with 12 industrial and research partners with the aim to improve the operational parameters for SWRO systems.

❑ Supervise the implementation of Greater Cairo rain drainage system whitch aims to construct a new system for rain drainage in the Greater Cairo area for the prevention of flooding due to short heavy rain falls.

❑ Supervise the implementation of the northern part of the Greater Cairo Ring road project

❑ Take part in the construction and project management of High rise buildings, special structures (marinas) and residential villas.

❑ Take part in the design of Cairo underground surface stations of phase 2

❑ The effect of aggregate impurities and synthetic fiber addition on high strength concrete failure behavior (graduation project).

Work Experience

2011 to date   Lead R&D Engineer, DOW chemical, Saudi Arabia.
www.DOW.com

2009 to 2011   Desalination and Water Treatment sector manager, Delft Environment, The Netherlands.
www.delft-environment.net

2007 to 2011   PhD researcher, TU Delft / UNESCO-IHE Institute, Delft, The Netherlands.
http://www.unesco-ihe.org

2004 to 2007   Senior plant sales and marketing Engineer, Metito Water Treatment, Egypt.
http://www.metito.com

2003 to 2004   Process research Engineer, PWN (The provincial water supply company of North Holland), The Netherlands.
http://www.pwn.nl

2002 – 2004   M.Sc. Study at UNESCO-IHE, Delft, The Netherlands
http://www.unesco-ihe.org

2001 – 2002 Infrastructure project engineer, Ministry of infrastructure, housing and new developed societies
http://www.housing-utilities.gov.eg

1999 – 2001 Senior Project Engineer/Junior project manager, Supply & Building LTD., Egypt

1997 – 1999 Civil design Engineer, BECT consultant, Egypt
http://www.BECT.net

Professional Organisations

- ❏ A member of the association of professional engineers in Egypt.
- ❏ A member in the European Desalination Society (EDS)
- ❏ A member in the International Desalination Association (IDA)

T - #0101 - 071024 - C184 - 244/170/10 - PB - 9780415615785 - Gloss Lamination